152 Topics in Current Chemistry

Electrochemistry IV

Editor: E. Steckhan

With contributions by
J. Heinze, A. Merz, H.-J. Schäfer

With 22 Figures and 21 Tables

Springer-Verlag Berlin Heidelberg GmbH

This series presents critical reviews of the present position and future trends in modern chemical research. It is addressed to all research and industrial chemists who wish to keep abreast of advances in their subject.

As a rule, contributions are specially commissioned. The editors and publishers will, however, always be pleased to receive suggestions and supplementary information. Papers are accepted for "Topics in Current Chemistry" in English.

DOI 10.1007/978-3-540-48139-3

This work is subject to copyright. All rights are reserved, whether the whole or part of the material is concerned, specifically the rights of translation, reprinting, re-use of illustrations, recitation, broadcasting, reproduction on microfilms or in other ways, and storage in data banks. Duplication of this publication or parts thereof is only permitted under the provisions of the German Copyright Law of September 9, 1965, in its version of June 24, 1985, and a copyright fee must always be paid. Violations fall under the prosecution act of the German Copyright Law.

© Springer-Verlag Berlin Heidelberg 1990
Originally published by Springer-Verlag Berlin Heidelberg New York in 1990
MyCopy version of the original edition 1990

The use of registered names, trademarks, etc. in this publication does not imply, even in the absence of a specific statement, that such names are exempt from the relevant protective laws and regulations and therefore free for general use.
www.springer.com/mycopy

Guest Editor

Professor Dr. *Eberhard Steckhan*
Institut für Organische Chemie und Biochemie,
Universität Bonn, Gerhard-Domagk-Str. 1, D-5300 Bonn 1

Editorial Board

Prof. Dr. *Michael J. S. Dewar*   Department of Chemistry, The University of Texas
Austin, TX 78712, USA

Prof. Dr. *Jack D. Dunitz*   Laboratorium für Organische Chemie der
Eidgenössischen Hochschule
Universitätsstraße 6/8, CH-8006 Zürich

Prof. Dr. *Klaus Hafner*   Institut für Organische Chemie der TH
Petersenstraße 15, D-6100 Darmstadt

Prof. Dr. *Edgar Heilbronner*   Physikalisch-Chemisches Institut der Universität
Klingelbergstraße 80, CH-4000 Basel

Prof. Dr. *Shô Itô*   Department of Chemistry, Tohoku University,
Sendai, Japan 980

Prof. Dr. *Jean-Marie Lehn*   Institut de Chimie, Université de Strasbourg, 1, rue
Blaise Pascal, B. P. Z 296/R8, F-67008 Strasbourg-Cedex

Prof. Dr. *Kurt Niedenzu*   University of Kentucky, College of Arts and Sciences
Department of Chemistry, Lexington, KY 40506, USA

Prof. Dr. *Kenneth N. Raymond*   Department of Chemistry, University of California,
Berkeley, California 94720, USA

Prof. Dr. *Charles W. Rees*   Hofmann Professor of Organic Chemistry, Department
of Chemistry, Imperial College of Science and Technology,
South Kensington, London SW7 2AY, England

Prof. Dr. *Fritz Vögtle*   Institut für Organische Chemie und Biochemie
der Universität, Gerhard-Domagk-Str. 1,
D-5300 Bonn 1

Preface to the Series on Electrochemistry

The scope of electrochemistry having broadened tremendously within the last ten years has become a remarkably diverse science. In the field of electroorganic synthesis, for example, selectivity has been improved by use of electrogenerated reagents, energy uptake lowered and space-time yields have been improved by using mediated reactions. In addition, electroorganic chemistry has been efficiently applied to the synthesis of key building blocks for complex molecules and has established its role as a new tool in organic synthesis. However electrochemistry has also found new and interesting applications in quite different fields of chemistry. Photoelectrochemistry, as one example, is not only valuable for transformations of organic molecules but also for the very important goal of energy conversion. More insight has been gained in the processes occurring on illuminated semiconductor electrodes and micro particles. Designing the composition of electrode surfaces can lead to the selective activation of electrodes. Electrochemical sensors and techniques present new opportunities for the analysis of biological compounds in medicine and biology. Research in the field of conducting polymers is very intensive because of interesting potential applications.

Therefore I am very happy that Springer-Verlag has decided to account for these important developments by introducing a series of volumes on new trends in electrochemistry within its series Topics in Current Chemistry. The volumes will cover the important trends in electrchemistry as outlined above in the following manner:

Electroorganic Synthesis by Indirect Electrochemical Methods;
New Applications of Electrochemical Techniques;
Recent Development in Electroorganic Synthesis.

The guest editor is very happy and thankful that well-known experts who are actively engaged in research in these fields have agreed to contribute to the volumes. It is hoped that this collection of reviews is not only valuable to investigators in the respective fields but also to many chemists who are not so familiar with electrochemistry.

Bonn, Mai 1987   Eberhard Steckhan

Preface to Volume IV

The fourth volume of the electrochemistry series in Topics in Current Chemistry combines three contributions from the field of organic electrochemistry. They demonstrate the broad scope of this research area which in itself is only a small part of electrochemical science.

One contribution belongs to the technically very important field of material science. It covers the exciting developments in the area of conducting polymers. It is reviewed from the standpoint of an electrochemist. Thus, besides giving an overview of the promising future applications, the formation of the polymers by electropolymerization, the charging/discharging properties, and the structures of conducting polymers are treated in detail.

The chemical modification of electrode surfaces is a field which has grown tremendously during the last 10 years. In most cases the modifying coatings are designed to contain electroactive groups. In this way, surface properties can be controlled by electrode potentials. A larg number of possible applications have therefore been developed. Among them are potential dependent color displays, sensors, molecular based electronic devices, catalysis of electrochemical reactions and many others. This article is mainly concerned with new techniques for the preparation of modified electrodes and with their application, in particular, in the field of synthetic and analytical organic electrochemistry. It is in some respects related to the first contribution to this volume because in some cases modifications of electrode surfaces use electrochemically generated conducting polymers as backbones.

The third contribution deals with an rather old synthetic electroorganic reaction, Kolbe-Electrolysis. However, recently this reaction has found so many interesting and important new applications, for example in the synthesis of naturally products or the generation of versatile, even enatiomerically pure, synthetic building blocks, that it was necessary to cover these new developments in a review. As these reactions are usually very easy to perform without expensive equipment, it is hoped that further applications in many laboratories will be initiated by this article.

Bonn, July.1989   Eberhard Steckhan

Table of Contents

Electronically Conducting Polymers
J. Heinze 1

Chemically Modified Electrodes
A. Merz 49

Recent Contributions of Kolbe Electrolysis to Organic Synthesis
H.-J. Schäfer 91

Author Index Volumes 151–152 153

Table of Contents of Volume 142

Electrochemistry I

Organic Syntheses with Electrochemically Regenerable Redox Systems
E. Steckhan

Selective Formation of Organic Compounds by Photoelectrosynthesis at Semiconductor Particles
M. A. Fox

Oxidation of Organic Compounds at the Nickel Hydroxide Electrode
H.-J. Schäfer

Electrogenerated Bases
J. H. P. Utley

The Chemistry of Electrogenerated Acids (EGA)
K. Uneyama

Table of Contents of Volume 143

Electrochemistry II

Electrochemical Techniques in Bioanalysis
C. E. Lunte, W. R. Heineman

Medical Applications of Electrochemical Sensors and Techniques
G. S. Calabrese, K. M. O'Connell

Photoelectrochemical Solar Energy Conversion
R. Memming

Mechanism of Reactions on Colloidal Microelectrodes and Size Quantization Effects
A. Henglein

Table of Contents of Volume 148

Electrochemistry III

Organic Electrosyntheses in Industry
D. Degner

Organic Electroreductions at Very Negative Potentials
E. Kariv-Miller, R. I. Pacut, G. K. Lehman

Synthesis of Alkaloidal Compounds Using an Electrochemical Reaction as a Key Step
T. Shono

Role of the Electrochemical Method in the Transformation of *beta*-Lactam Antibiotics and Terpenoids
S. Torii, H. Tanaka, T. Inokuchi

Electronically Conducting Polymers

Jürgen Heinze

Institut für Physikalische Chemie der Universität Freiburg, Albertstr. 21, D-7800 Freiburg, FRG

1 Introduction	2
2 Electropolymerization	3
2.1 Historical Background	3
2.2 Mechanisms of Electropolymerization	6
2.3 The Electrodeposition Process	13
3 Structure of Conducting Polymers	16
4 Charge Storage Mechanism in Conducting Polymers	17
5 Applications	29
5.1 Rechargeable Batteries	30
5.2 Electrochromic Devices	33
5.3 Miscellaneous	33
6 Acknowledgements	35
7 Note Added in Proof	36
8 References	38

The exciting new field of conducting polymers has been reviewed from the viewpoint of electrochemistry. Their connection to all kinds of modern electrochemical research results from their unusual charging/discharging properties. In addition, electropolymerization is a challenging method of preparing these materials. Therefore, main topics treated are electropolymerization, the structure, and the charge storage mechanism of conducting polymers. An overview of prospective applications is given.

Jürgen Heinze

1 Introduction

Over the past 15 years the physics and chemistry of organic metals have developed into a rich field of interdisciplinary research with ever-widening perspectives, both with respect to basic research and technical application. Initially, interest was concentrated on organic donor-acceptor complexes [1-5]. This changed in 1977 with the discovery by the American scientists Heeger and MacDiarmid that doping polyacetylene (PA) with iodine endowed the polymer with metallic properties, including an increase in conductivity of 10 orders of magnitude [6,7]. The successful doping of PA — in electrochemical terminology the equivalent of oxidation or reduction — encouraged the same scientists to test PA as a rechargeable active battery electrode [8,9]. Their promising results triggered off world-wide efforts to construct a polymer battery. In the course of these studies conducting polymers with properties similar to PA were discovered, such as polypyrrole (PPy) and polythiophene (PTh). However, it soon became apparent that the pathway from inspiration to practical realization can be long and thorny, and that basic research on a broad, interdisciplinary basis is necessary to reach the goal.

PA ...

PPP ...

PPy ...

PTh ...

PANI ...

It was also observed that, with the exception of polyacetylene, all important conducting polymers can be electrochemically produced by anodic oxidation; moreover, in contrast to chemical methods, the conducting films are formed directly on the electrode. This stimulated research teams in the field of electrochemistry to study the electrosynthesis of these materials. Most recently, new fields of application, ranging from anti-corrosives through modified electrodes to microelectronic devices, have aroused electrochemists' interest in this class of compounds [10-12].

A structural characteristic of conducting organic polymers is the conjugation of the chain-linked electroactive monomeric units, i.e. the monomers interact via a π-electron system. In this respect they are fundamentally different from redox polymers. Although redox polymers also contain electroactive groups, the polymer backbone is not conjugated. Consequently, and irrespective of their charge state, redox polymers are nonconductors. Their importance for electrochemistry lies mainly in their use as materials for modified electrodes. Redox polymers have been discussed in depth in the literature [13-15], and will not be included in this review.

Given the different intentions and objectives of research into conducting polymers, it is inevitable that the relevant literature is widely scattered over numerous journals, and that some of it, especially research abstracts and proceedings of meetings [16-22], is difficult to find. Accordingly, this review will primarily take stock of the electrochemical literature on the subject of conducting polymers and point out current trends in research. The large body of theoretical and experimental work on the characterization of the electrical, magnetic, and optic properties of these materials will receive only brief attention; the relevant information can be found in a number of review articles and monographs published in recent years [23-30].

2 Electropolymerization

2.1 Historical Background

The existence of materials now included among the conducting polymers has long been known. The first electrochemical syntheses and their characterization as insoluble systems took place well over a century ago. In 1862 Letheby [34] reported the anodic oxidation of aniline in a solution of diluted sulphuric acid, and that the blue-black, shiny powder deposited on a platinum electrode was insoluble in H_2O, alcohol, and other organic solvents. Further experiments, including analytical studies, led Goppelsroeder [35] to postulate in 1876 that "oligomers" were formed by the oxidation of aniline.

In 1968 Dall'Olio et al. [36] published the first report of analogous electrosyntheses in other systems. They had observed the formation of brittle, filmlike pyrrole black on a Pt-electrode during the anodic oxidation of pyrrole in dilute sulphuric acid. Conductivity measurements carried out on the isolated solid state materials gave a value of 8 Scm^{-1}. In addition, a strong ESR signal was evidence of a high number of unpaired spins. Earlier, in 1961, H. Lund [37] had reported — in a virtually unobtainable publication — that PPy can be produced by electrochemical polymerization.

In 1979, Diaz et al. [38] produced the first flexible, stable polypyrrole (PPy) film with high conductivity (100 Scm^{-1}). The substance was polymerized on a Pt-electrode by anodic oxidation in acetonitrile. The then known chemical methods of synthesis [39-41] usually produced low conductivity powders from the monomers. By contrast, electropolymerization in organic solvents formed smooth and manageable films of good conductivity. Thus, this technique soon gained general currency, stimulating further electropolymerization experiments with other monomers. In 1982, Tourillon

Table 1. Conducting polymers: preparation methods and conductivities

Name	Structure	Main preparation method[a]	Conductivity[b] [S/cm]	Refs.
Polyacetylene		Ch	100 >100000	6) 254)
Polypyrrole		EP	100	38)
Polythiophene		EP	106	42)
Polyfuran		EP		42)
Polyindol		Ch		42)
Polycarbazole		EP	10^{-3}–10^{-4}	49)
Polyparaphenylene		Ch/EP	500	50, 51)
Polyaniline		EP		52)
Polyazulene		EP	10^{-1}–1	49)
Polynaphthalene		EP	10^{-3}	47)
Polyanthracene		EP	10^{-3}	48)
Polypyrene		EP	10^{-1}–1	49)

Table 1. (continued)

Name	Structure	Main preparation method[a]	Conductivity[b] [S/cm]	Refs.
Polydithienothiophene		EP	10^{-2}	56)
Polythieno[3,2-b]pyrrole		EP	5×10^{-3}	57)
Poly[1,4-di-(2-thienyl)benzene]			1×10^{-4}	58)
Poly(3-alkylthienylene)		EP	5×10^{-3}	57)
Polyparaphenylene-vinylene		Ch	0.1–2800	60, 152)
Polyimides		Ch		61)
Polythiazyl		Ch	10^3	62)
Polyquinolines		Ch	10^{-4}	264)
Polyphenylenesulfide		Ch	7	311, 312)
Polythieno[3,4-c]thiophene		unknown	—	313)
Polythienylenevinylene		Ch	37–62	314)
Polyfuranylvinylene		Ch	4–36	314)

Table 1. Conducting polymers: preparation methods and conductivities

Name	Structure	Main preparation method[a]	Conductivity[b] [S/cm]	Refs.
Polyfluorene		EP	10^{-4}	53, 54)
Polyisothianaphthalene			50	55)

[a] Ch = Chemical Preparation; EP = Electropolymerization;
[b] in the doped state

and Garnier [42] showed that anodic oxidation of hetero- and homocyclic monomers produced polythiophene (PTh), polyfuran (PFu), polyindole (PIn) and polyazulene (PAz). Independent of these experiments, two research teams reported the electropolymerization of benzene to polyparaphenylene (PPP) in SO_2/Me_4NBF_4 and HF/SbF_5 [43,44]. However, in the latter instances the film quality was unsatisfactory and the conductivity values did not exceed 10^{-3} Scm^{-1}. Since then, this principle of synthesis has been used to polymerize numerous substituted heterocyclic compounds of the pyrrole and thiophene groups as well as other parent substances [45]. Recently, alternative electrochemical methods based on a Ni-catalyzed reductive polymerization or a Cu-catalyzed oxidative polymerization have enable the production of even (PA) [46] and other barely accessible polyaromatics as conductive films on electrodes [47,48]. The number of conducting polymer systems is still increasing. Furthermore, structural modifications, among other things, have resulted in qualitatively new properties such as solubility [27]. Table 1 presents all the important conducting polymers known at present.

2.2 Mechanisms of Electropolymerization

The most widely used method of producing intrinsically conductive polymers is the anodic oxidation of suitable monomer species. This electrosynthesis has two particular advantages: the polymers are formed in the doped, i.e. conducting state, and the films, as a rule, posses favourable mechanical properties. In the usual electrochemically induced polymerization reactions [63], the electrode catalytically triggers chain growth and, consequently, the process requires little electricity. By contrast, the anodic oxidation has an electrochemical stoichiometry of 2.07 to 2.7 faraday/mol of reacting monomer. It has been deduced from numerous measurements of the electropolymerization of PPy and PTh as well as from elementary analysis [64-66] that the film-forming process needs only 2 faradays/mol, i.e. 2 electrons/molecule, and the additional charge serves the partial, reversible oxidation (doping) of the polymer film. As the potential needed for monomeric oxidation is always significantly higher than the charging of the existing polymer, the two processes — film formation and its oxidation — occur simultaneously. The first spectroscopic studies of PPy [67] and other systems [42, 68], as well as oxidative decomposition experiments with PPy — which produce mainly pyrrole-2.5-dicarboxylic acids [40] — had already shown that the basic structure of the monomeric subunit survives in the polymer. Polymerization experiments with various substituted pyrrole monomers showed further that only α-substituted derivatives did not polymerize [69]. From these results one concluded that the polymerization of pyrrole and thiophene via cationic intermediates leads to linear polymers whose monomeric units were α,α'-bonded. As in this process protons in the α-position are eliminated [70], this polymerization process is classified as a condensation reaction.

Taking the electrochemical stoichiometry into account, the complete reaction equation for the polymerization of a species HMH is:

$$(n + 2)\, HMH \rightarrow HM(M)_n\, MH^{(nx)+} + (2n + 2)\, H^+ +$$
$$+ (2n + 2 + nx)\, e^-. \qquad (1)$$

$(2n + 2)$ electrons are used for the polymerization reaction itself, while the additional charging of the polymer film requires nx electrons. In general x lies between 0.25 and 0.4; this means that every 3rd to 4th monomeric subunit in the film is charged.

The mechanism of electropolymerization, and in particular the initiation of the process, has still not been completely explained. The one certainty is that in the very first step the neutral monomer is oxidized to a radical cation. By analogy to the longknown coupling reactions of radical cations in aromatic compounds [71, 72], Diaz et al. have suggested that in the polymerization of pyrrole the monomers dimerize (radical-radical coupling = RRC) after oxidation at the electrode, and that protons are eliminated from the doubly-charged dihydrodimer, forming the neutral species [64, 73]. As the dimer (E_{pa} = 0.6 V vs. SSCE) is more easily oxidizable than the monomer (E_{pa} = 1.2 V vs. SSCE), under the given experimental conditions it is immediately reoxidized to the cation. Chain growth is accompanied by the addition of new cations of the monomeric pyrrole to the already charged oligomers (dimers). This in turn is followed by another proton elimination and the oxidation of the propagated oligomeric unit to a cation (Scheme I: a → b → c → f → g). In the terminology of electrochemical reaction mechanisms, this chain propagation process corresponds to a cascade of ECCE steps (E = *E*lectron transfer, C = *C*hemical step).

Some authors [74-76] have questioned this mechanism on the ground that the strong coulomb repulsion between small cation radicals renders a direct dimerization of such particles improbable. Instead, they postulate a radical substrate coupling (= RSC). An electrophilic attack by the radical cation on the neutral monomer produces the single-charged dihydroderivative, which eliminates its protons only after a further charge transfer, becoming the neutral dimer (Scheme I: a → d → e → c ...). However, this suggestion is insufficient to characterize the chain propagation process as growth is observed experimentally only when the oxidation of the monomer occurs parallel to the oxidation of the film [77]. This also holds when the parent compound is already present as a dimer or a higher oligomer. This implies that the polymerization process following on the dimerization of the monomers must involve a change of mechanism.

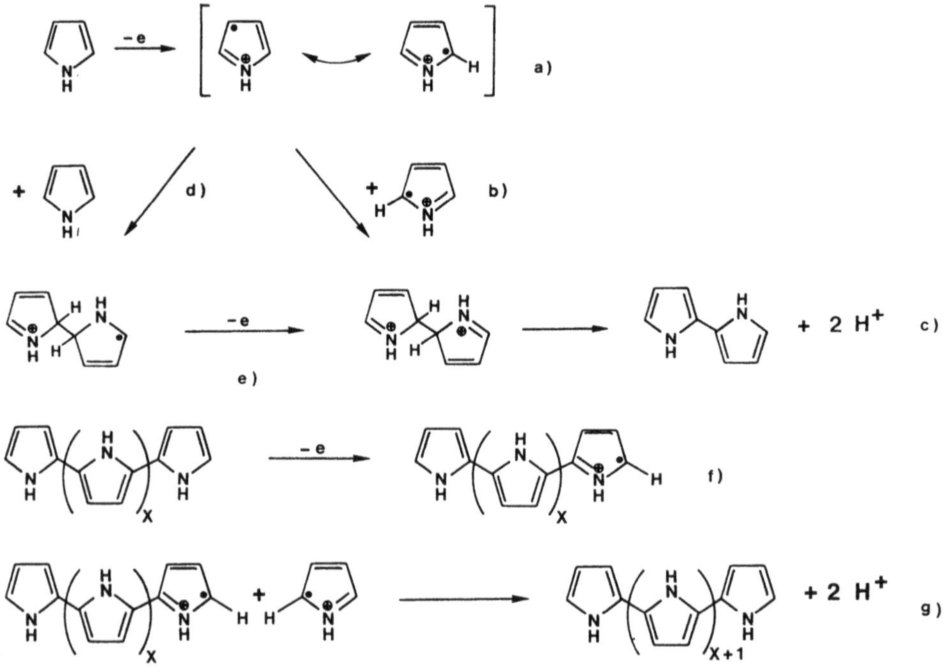

Scheme I

So far, only the electrocrystallization of radical cation salts, such as fluoranthene [78], has produced concrete evidence for the RS route. In these cases it leads to the formation of one-dimensional conductors of stack-like structure. The interaction within the primary product, the RS associate, is based solely on a charge-transfer effect, without any covalent bonding. There have been several recent reports [79, 80] on kinetic effects, which in the opinion of the authors, indicate the intermediate formation of a covalently bonded radical-substrate associate from anodically oxidized aromatics. However, the subsequent controversial discussion [81] has shown that consistent proof is still lacking. On the other hand, the principal objection to the RR reaction path — the strong coulomb repulsion between charged particles — is not convincing for reactions in solution. Using the Debye-Smoluchovski Theory [83], Eigen [82] had already shown that small particles with the same charge are able to associate at diffusion-controlled rate.

The difficulties in unequivocally characterizing the electropolymerization reactions of pyrrole or thiophene are largely a consequence of the high reaction rates, which make it impossible to analyse their kinetics within the available time scales of the electrochemical techniques. Owing to its slow coupling reaction compared to those of other polymerizable systems, electrochemical studies on the anodic oxidation of N-phenylcarbazole have provided very clear and convincing indications for the initiation steps of electropolymerization [84]. Although the electro-oxidation of carbazoles has frequently been studied [85–87] in the past, polymerizations have by and large been disregarded as disturbing side-effects. Cyclic voltammetric measurements on a concentrated solution (1.7×10^{-2} mol/l) of N-phenylcarbazole show that poly-

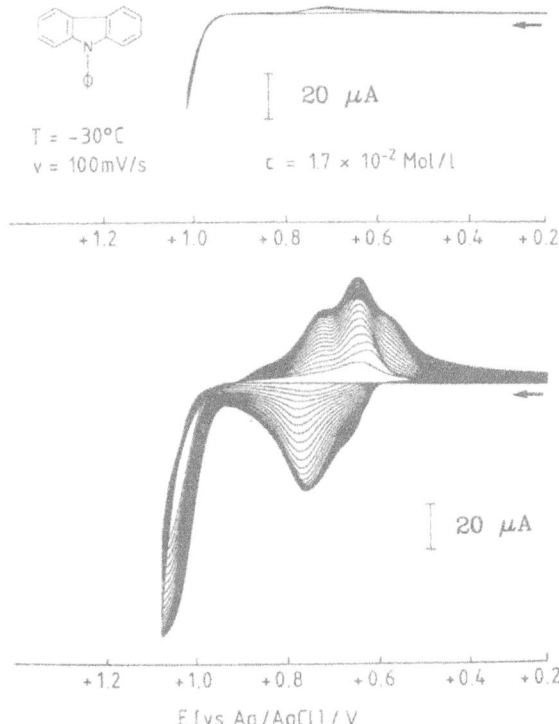

Fig. 1. Multisweep cyclic voltammograms for the polymerization of N-phenyl-carbazole ($C = 1.7 \times 10^{-2}$ mol/l) in CH_3NO_2/$TBAPF_6$, Pt working electrode. The switching potential in the upper diagram is 50 mV less positive than in the lower

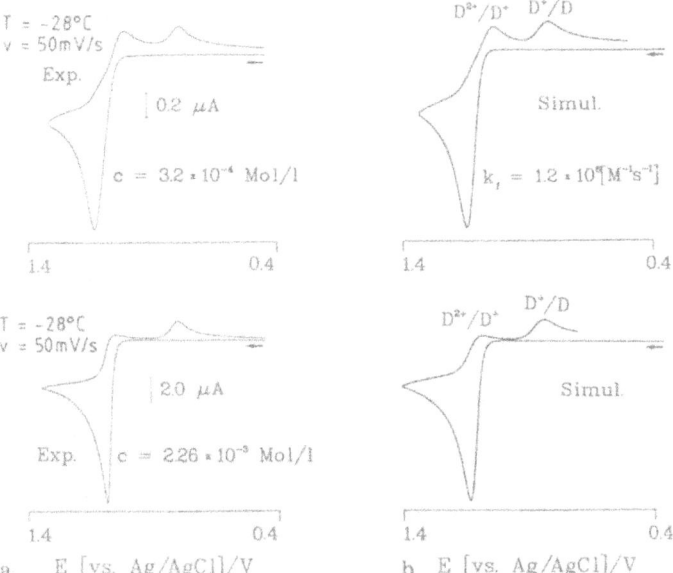

Fig. 2. Experimental (a) and simulated (b) voltammograms characterizing the beginning of the electropolymerization of N-phenylcarbazole (CH_3NO_2, Pt-electrode)

merization starts at a formation potential well before the thermodynamic redox potential of the monomer species (Fig. 1).

One also obtains analogous findings with trace-crossing effects[76] for the electropolymerization of thiophene and pyrrole. This cannot be explained by a simple linear reaction sequence, as presented in Scheme I, because it indicates competing homogeneous and heterogeneous electron transfer processes. Measurements carried out in a diluted solution of N-phenylcarbazole provide a more accurate insight into the reaction mechanism (Fig. 2).

Detailed kinetic studies in connection with digital simulations do confirm the RR coupling mechanism postulated in older publications [85-87] as well as the oxidation of the resulting dimer D to the dication D^{2+}. But the surprising drop in the height of the reduction wave for the D^{2+}/D^+ redox pair as the concentration of the parent monomer M increases proves that in solution, in addition to the heterogeneous process, a homogeneous redox comproportionation from D^{2+} with M to D^+ and M^+ takes place (Scheme II).

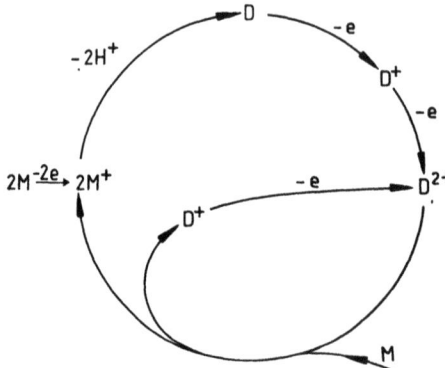

Scheme II

The dimerization continually disturbs the comproportionation equilibrium by withdrawing the monomer M^+. Consequently, M^+ is formed homogeneously in the diffusion layer, despite unfavourable thermodynamic conditions. The propagation process, too, is marked by analogous processes, in which M^+ is coupled to the polymer chain. The evaluation of the measured data give a dimerization constant k_D of 1.2×10^6 mol/ls for the very 'slow' dimerization reaction of the N-phenylcarbazole. Accordingly, the k_D values for pyrrole and thiophene must be greater by some orders of magnitude.

Independently of this, chronoabsorptiometric measurements by Genies et al. [73] have proved that PPy films grow in timer linear to t and not to $\sqrt{t}$. In the opinion of the authors this implies that the rate-determining step during film growth is a radical ion coupling and not the diffusion of the uncharged monomer towards the electrode surface. The attested phenomenon [38,64,88-90] that PPy polymerizes considerably faster in the presence of polar H_2O ($\varepsilon = 79$) than in dry aprotic solvents must be taken as further proof of a purely ionic RR coupling. On account of its high dielectric constant, water obviously reduces the coulomb repulsion between the charged pyrrole monomers and oligomers. This also explains why considerably more soluble oxidation products of pyrrole are formed in anhydrous CH_3CN than in aqueous CH_3CN (1 mol %) [102]. The observation of Lundström et al. [91], viz. that the polymerization rate of pyrrole significantly increases as the supporting electrolyte concentration of $TEABF_4$ in acetonitrile increases, points in the same direction. However, their contention that the oxidation of the BF_4^- ion to $BF_4^{\cdot} + e$ initiates the polymerization is not convincing as BF_4^- decomposes only above $+3.0$ V vs. SCE [76]. The fact that the tendency of thiophene to polymerize in CH_3CN is drastically reduced in the presence of small amounts of water appears contradictory. Downard and Pletcher's [92] potential step experiments indicated conspicuously low oxidation currents, from which they concluded that in the presence of water a nonconducting passivating layer forms on the Pt-electrode or on the PTh-surface. However, it is still unclear just how water accounts for the changes in the reaction mechanism between pyrrole and thiophene. A similar passivating effect occurs when benzene is oxidized in superdry acetonitrile [93]. But if dry SO_2 is used instead, adhesive polyparaphenylene films with suitable mechanical properties are formed [94].

These apparent contradictions can be resolved if one keeps in mind that the competition between several reaction paths is dependent upon both the reactivities of the anodically oxidized parent species *and* the polymer film as well as the reactivity of the surrounding solvent-electrolyte medium.

As a general rule, the polymerization rate rises with increasing oxidation potential in the order Py < Th < Fu < Bz. But this is qualified by nucleophilic solvents competing with the radical cation intermediate to couple with the growing polymer. The nucleophilicity of the solvents commonly used in electropolymerization rises in the following order: $HF < CF_3COOH < SO_2 < CH_3NO_2 < CH_2Cl_2 < PC < CH_3CN < H_2O$. In solvents up to and including SO_2 all the aforementioned monomers polymerize at the appropriate electrode potentials virtually without any secondary reactions. In the case of pyrrole, however, the rate constants of dimerization and oligomerization are markedly lower than in the case of benzene; accordingly, correspondingly high concentrations of the parent monomer are required to produce good polymer films. By contrast, benzene can be only partially polymerized in CH_3CN as the newly formed oligomers react with the solvent more readily than with the benzene radical cations, thereby blocking further growth. Consequently, only a passivating adsorbed layer forms on the electrode. Only pyrrole can be successfully polymerized in water [95-98]. The high DC of H_2O ($\varepsilon = 79$) accelerates the coupling reaction, while at the same time because of the low oxidation potential the nucleophilic reactivity between water and the polymer is diminished. Even in the case of thiophene, which, on account of its higher oxidation potential, is considerably more reactive than pyrrole and, therefore, easily polymerizes in CH_3CN, small amounts of water are sufficient for nucleophilic addition to block further growth of the oligomer chain.

Another complication arises from the little regarded circumstance that the reactivity of the growing polymer chain itself changes, and particularly in the initial growth phase. In contrast to the normal polycondensation, the reactivity of the chain is dependent on its charging, which, in turn, is a function of the electrode potential and the respective degree of polymerization. Hence, as a new chain begins to form the coupling rate falls 'rapidly', until it reaches a steady state characterized by a constant number of charges per segment despite an increasing degree of polymerization. Parallel to chain growth — during which monomers and dimers as well as lower oligomers are added to the existing polymer, particularly at high potentials — dimerization, trimerization and higher oligomerization takes place in solution in the diffusion layer of the electrode. The oligomers forming in the vicinity of the electrode couple with growing chains at different rates, depending on size, or are incorporated in to the polymer matrix, where, as charged nuclei, they can trigger off new associations. It is self-evident that the number of such competing parallel reactions rises with increasing positive potential, causing a broad molecular weight distribution and the formation of irregularly structured materials [77].

The above discussion on the formation mechanism of conducting polymers indicates the complexity of the reaction sequence. Hence, it is not surprising that there has been no lack of attempts to establish correlations between the tendency to polymerize and the physical properties of the starting monomers by systematically varying individual parameters. Unsubstituted monomers such as pyrrole and thiophene are particularly popular candidates for studying the effect of substitution on polymerizability as well as other related properties. Apart from steric effects, the influences of substitution on electron behaviour have received special attention. Experiments performed by Salmon et al. [100] with N-substituted pyrroles have shown that sterically crowded substituents, such as the *tert*-butyl- or the cyclohexyl-group, totally block polymeri-

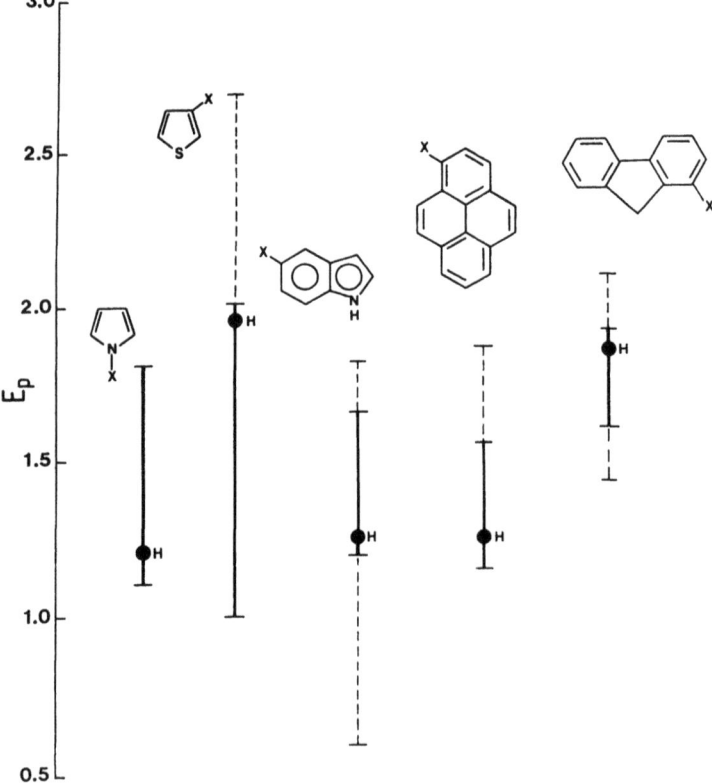

Fig. 3. Accessible potential ranges for the electropolymerization of substituted pyrroles, thiophenes, indoles, pyrenes and fluorenes

zation, whereas long alkyl chains [101], halogene substituents [102] or *ortho*-substituents on the phenyl ring [103] produce films with poor conducting and mechanical properties. In contrast to the steric effects, the purely electronic influences of substituents are less clear. They are best documented by linear free-energy relationships, which, for the cases in question, are for the most part only plots of voltammetrically obtained peak oxidation potentials of corresponding monomers against their respective Hammett substituent constant [104]. As a rule, the linear correlations are very good for all systems, and prove, in accordance with the Hammett-Taft equation, the dominance of electronic effects in the primary oxidation step. But the effects of identical substituents on the respective system's tendency to polymerize differ from parent monomer to parent monomer. Whereas thiophenes which receive electron-withdrawing substituents in the, as such, favourable β-position do not polymerize at all [66], indoles with the same substituents polymerize particularly well [105].

This contradictory behaviour is reflected in the Hammett plots by a high slope value $\varrho = 0.80$ for the thiophenes and a lower $\varrho = 0.56$ for the indoles. Waltmann, Diaz and Bargon [105,106] explain this in terms of differences in the expansion of the π-electron systems in the two unsubstituted monomers. Accordingly, there are differences in the stability of the corresponding radical cations. Consequently, for each substance class there is an optimal reaction range for electropolymerization which depends on the substituents present. In general, taking the Hammett plots together with the available polymerization data, this means that parent substances with a high oxidation potential will preferably form polymers with electron-donating substituents, and those with a low oxidation potential with electron-withdrawing substituents (Fig. 3).

The intermediate radical cations formed at the electrode have in principle a choice of three reaction paths [105,106]. These are, first, the polymerization reaction (k_p), second, in the event of high cationic stability, diffusion into the solution (k_d) and, third, for very reactive particles, reaction with the solvent or the anions of the supporting electrolyte at the electrode interface ($k_n([solv] + [X^-])$). The fraction f_p of the radical cations which electropolymerize is then given by the equation

$$f_p = \frac{k_p}{k_p + k_d + k_n([solv] + [X^-])} \qquad (2)$$

Thus, polymerization will always occur when $k_p \gg k_d + k_n([solv] + [X^-])$. In the case of highly electron-donating substituents the stability of the radical cations may be so great that $k_d \gg k_p + k_n([solv] + [X^-])$ and most of the ions diffuse into the solution. By contrast, if — given electron-withdrawing subsituents and high oxidation potential — $k_n([solv] + [X^-])$ becomes greater than $k_p + k_d$, then the nucleophilic addition will dominate and the polymerization will be suppressed.

2.3 The Electrodeposition Process

Despite the vast quantity of data on the chemistry of electropolymerization, relatively little is known about the processes involved in the deposition of polymers on the electrode, i.e. the heterogeneous phase transition. Research — voltammetric and potential step experiments — has largely concentrated on the induction stage of film formation of PPy [76,107–109], PTh [92,110] and PANI [111].

A trace-crossing appears on the reverse sweep of the first cycle in all voltammograms, providing that the scan reversal lies close to the peak potential.

Such effects are observed inter alia [112] when a metal is electrochemically deposited on a foreign substrate (e.g. Pb^{2+} on graphite), a process which requires an additional nucleation overpotential. Thus, in cyclic voltammetry metal is deposited during the reverse scan on an identical metallic surface at thermodynamically favourable potentials, i.e. at positive values relative to the nucleation overpotential. This generates the typical trace-crossing in the current-voltage curve. Hence, Pletcher et al. [76,107] also view the trace-crossing as proof of the start of the nucleation process of the polymer film, especially as it appears only in experiments with freshly polished electrodes. But this is about as far as we can go with cyclic voltammetry alone. It must be complemented by other techniques; the potential step methods and optical spectroscopy have proved suitable.

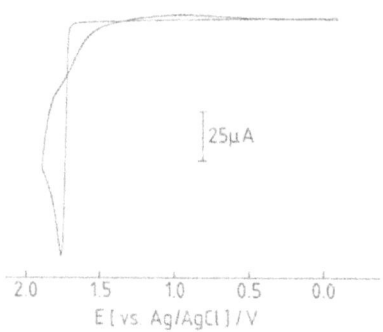

Fig. 4. First voltammetric cycle of the oxidation of thiophene in CH_3CN measured with a freshly polished Pt-electrode

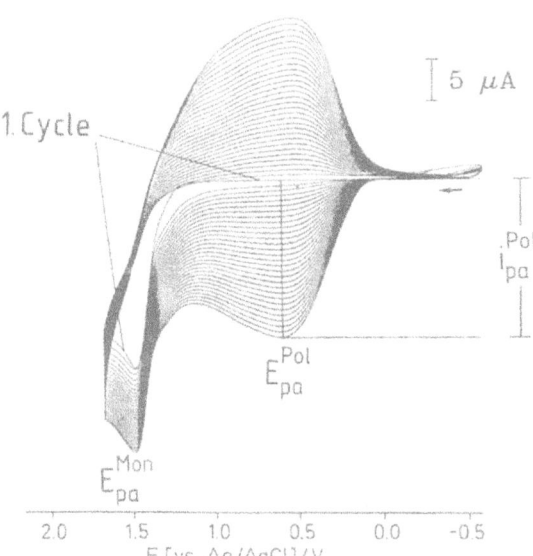

Fig. 5. Potentiodynamic growth of a polypyrrole film in wet accetonitrile (v = 0.2 V/s, T = 24 °C, number of cycles n = 30); E_{pa}^{mon} = +1.50 V vs. Ag/AgCl, E_{pa}^{pol} = +0.62 vs, Ag/AgCl

Galvanostatic, potentiostatic as well as potentiodynamic techniques can be used to electropolymerize suitable monomeric species and form the corresponding film on the electrode. Provided that the maximum formation potentials for all three techniques are the same, the resulting porperties of the films will be broadly similar. The potentiodynamic experiment in particular provides useful information on the growth rate of conducting polymers. The increase in current with each cycle of a multisweep CV is a direct measure of the increase in the surface of the redoxactive polymer and, hence, a suitable measure of relative growth rates (Fig. 5).

The relative growth rate per cycle v is calculated from the peak current of the respective polymer oxidation using Eq. (3):

$$v = k \cdot i_{pa}^{pol}/(n-1) \qquad (3)$$

where k is a proportionality constant. The parameter v is by its nature a function of the given conditions of polymerization and, thus, must be standardized to v_0, in relation to which only one growth parameter may be changed [77b].

Costa and Garnier [113] conclude from optical measurements based on the differential ellipsometry of 2-methylthiophene in CH_3CN that the intermediate radical cations form oligomers in coupling reactions in the solution at the electrode interface with a degree of polymerization of 3 to 4. Above a critical chain length insoluble oligomers form, which deposit themselves on the electrode surface. Two findings support this thesis. In the potential step experiment the optical signal indicating the deposition of oligomers on the electrode only changes when the current on the i–t curve starts to rise again after the typical $t^{-1/2}$ drop. In the potentiodynamic experiment the observed oxidation currents, as well as the visible changes on the electrode surface, are dependent on the scan rate. The slower the scan rate, the more visible the electrode coating and the greater the rise in the oxidation current. These results speak against the assumption [76,110] of primary adsorption of monomers on the electrode surface, followed by oxidation and polymerization in the adsorbed state to form a single polymeric layer on the electrode.

In i–t experiments with thiophene, pyrrole and aniline monomers the observed law of phase formation is usually $i \sim t^2$, after induction periods of varying length with a $t^{-1/2}$ characteristic. In most publications [107-111] this relationship is interpreted as instantaneous nucleation accompanied by three-dimensional growth of hemispheres [114]. Depending on the monomeric concentration in the solution, the high density of the nuclei lead more or less rapidly to mutual overlapping of the hemispheres, so that a film can continue to grow only perpendicularly to the surface. Although the three-dimensional growth model — sometimes the discussion also refers to the two-dimensional progressive nucleation of discs — is generally favoured on the basis of experimental data, all models of phase formation are viewed with considerable reservation. The objections to the generally accepted presentation are two-fold. Firstly, the models developed by Fleischmann and Thirsk [114] as well as other authors [115] are based on nucleation and growth of metallic phases and anodically deposited oxide films. The chemistry of the elementary steps of these processes, and thus their charging, differ fundamentally from the processes in the electropolymerization of organic materials. Secondly, a prerequisite for the postulated three-dimensional growth is chain-branching reactions, for which there is no support in the available data on structure (see Sect. 3).

3 Structure of Conducting Polymers

Besides synthesis, current basic research on conducting polymers is concentrated on structural analysis. Structural parameters — e.g. regularity and homogeneity of chain structures, but also chain length — play an important role in our understanding of the properties of such materials. Research on electropolymerized polymers has concentrated on polypyrrole and polythiophene in particular and, more recently, on polyaniline as well, while of the chemically produced materials polyacetylene still attracts greatest interest. Spectroscopic methods have proved particularly suitable for characterizing structural properties [116]. These comprise surface techniques such as XPS, AES or ATR, on the one hand, and the usual methods of structural analysis, such as NMR, ESR and X-ray diffraction techniques, on the other hand.

PPy was the first conducting polymer to be structurally analyzed. The discovery that α,α'-disubstituted pyrroles did not electropolymerize led to the conclusion that the pyrrole units in PPy are α-linked [69,117]. Magic angle spinning ^{13}C-NMR data [67,118] support the view that the pyrrole units bond chiefly in the α,α'-position, although α,β bondings are also found. Further support is given by IR measurements [67,115], which show very similar spectra for polypyrrole films and the α,α'-bonded pyrrole trimer. On the other hand, XPS measurements of polypyrrole [119] reveal that about one-third of the pyrrole rings in a chain are irregularly bonded. In contrast, as evidenced by XPS spectra, the structure of polymers from β,β'-dimethylpyrrole is by and large regular, for, in this case, polymerization is possible only in the α-position. Although X-ray structural analyses of neutral PPy provide only very general information — the material is not very crystalline — this, too, supports the assumption of a linear chain structure in which the orientation of the α,α'-bonded pyrrole molecules alternates [120]. The elementary cell is assumed to be a monoclinic unit, an assumption based not on experimental results but deduced from models.

Similarly, a chain structure with predominantly α,α'-coupling between the monomer units is postulated for polythiophene (PTh) on the basis of spectroscopic findings. Especially IR measurements of uncharged, chemically produced samples clearly reveal that the aromatic parent unit is not changed by polymerization [121]. The (C—H) valence vibration at 788 cm^{-1} is characteristic for the β-position of the thiophene ring and proves that α,α'-bonding predominates in the polymers produced from α,α'-dibromothiophene [68,42]. By contrast, specially synthesized poly-(α,β)-thiophene has two characteristic (C—H) vibrations at 730 and 820 cm^{-1} [122]. The analysis of the C—H stretching vibrations for electrochemically produced PTh indicates both regular α,α'-bonded PTh as well as disordered α,β bonded segments. An additional band at 1682 cm^{-1} also indicates defects in the polymer chain, corresponding inter alia to C=O bondings, caused by reactions with H_2O or O_2 [123]. On the other hand, as the IR findings prove, the electropolymerization of 2,2'-bithiophene or of 3-methylthiophene produces very regular, homogeneous materials, in which the monomeric units are almost exclusively α,α'-connected [124,125]. The IR data correlate very well with ^{13}C-NMR measurements, which similarly confirm the dominance of α,α'-bonding in P3-MeTh [126].

In contrast to the classic conducting polymers such as PPy, PTh, PP or PA, structural analyses of other systems are few and far between and limited for the most part to quantum mechanical model calculations on the formation of an ideal polymer structure. Open-shell INDO calculations of, above all, Diaz, Bargon and Waltman [105,106] have determined the positions of greatest spin density for some aromatic and heterocyclic radical cations. From PPy studies it is known that these correspond to the positions of greatest reactivity for a radical-ion coupling. The calculations correlate with the experimental findings insofar as monomers whose carbon atoms of high spin density have been substituted do not polymerize.

Chain length is another factor closely related to the structural characterization of conducting polymers. The importance of this parameter lies in its considerable influence on the electric as well as the electrochemical properties of conducting polymers. However, the molecular weight techniques normally used in polymer chemistry cannot be employed on account of the extreme insolubility of the materials. A comparison between spectroscopic findings (XPS, UPS, EES) for PPy and model calculations has led some researchers to conclude that 10 is the minimum number of monomeric units in a PPy chain, with the maximum within one order of magnitude [119,127,128]. But the mechanical qualities of the electropolymerized films, which are at times superb, speak against this conjecture. By electropolymerizing α,α'-tritium labeled β,β-dimethylpyrrole and comparing the tritium activity in the monomer and the polymer, Nazzal and Street [129] obtained molecular weights which indicate chain lengths of between 100 and 1000 pyrrole units. Sato et al. [130] determined a very much lower degree of polymerization, viz. between 10 and 20 units, for the Ni-catalyzed polycondensation of naphthalene dibromides, under the assumption that the polymer retained the bromine substituents at the respective chain ends. Recently, the molecular weight of polymers was directly measured for the first time. Soluble poly(3-alkyl)-thiophenes electropolymerized from the corresponding monomers were analyzed using an HPLC technique [131,132]. Their molecular weights M lay between 20000 and 40000, which corresponds to approximately 150 to 300 monomeric units. Recently, L. L. Miller et al. [133] found astonishingly short chains of maximum 10 monomers for the soluble poly(3-methoxy)thiophenes. The mole mass was determined by gel permation chromotography (GPC) of reduced material, with polystyrene as the standard of comparison.

4 Charge Storage Mechanism in Conducting Polymers

Conducting polymers, provided they are chemically produced, are initially insulators. Their metal-like properties, i.e. their high conductivity and optic reflectivity, only become obvious after "doping". Even in the earliest stages of research on these materials, it was clear that these processes were not comparable with the classical doping of typical semiconductors. Rather, they correspond to oxidation in the case of p-doping or reduction in the case of n-doping. Suitable redox reagents are either chemical electron acceptors, such as iodine, or electron donors, such as potassium naphthalide; or the process may be electrochemically induced via an electrochemical cell. Because of the redox reaction the polymer chain is negatively charged in the case of reduction and positively charged in the case of oxidation. To maintain electroneutrality the appropriate counterions diffuse into the polymer during charging and out of the polymer during discharging.

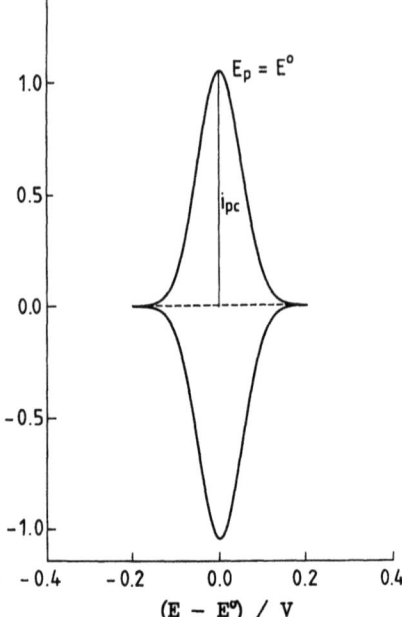

Fig. 6. Theoretical cyclic voltammogram for a redox-active polymer film with noninteracting redox centres

The knowledge that conducting polymers can be charged, i.e. oxidized and reduced, raised early on the question of possible applications, such as the construction of a polymer battery. But basic research was long unable to explain the charge storage mechanism.

One major reason for the diversity of views on the redox states of conducting polymers is the variety of possible shapes and forms of potentiodynamic current-voltage curves, even when the materials are prepared under more or less 'similar' conditions.

In the ideal case, reversible cyclic voltammograms of redoxactive films should show completely symmetrical and mirror-image cathodic and anodic waves with identical peak potentials and current levels [134–137] (Fig. 6).

The current in the reversible case is then:

$$i = \frac{n^2 F^2 A \Gamma_T v \cdot \exp \Theta}{RT(1 + \exp \Theta)^2}, \tag{4}$$

where

$$\Theta = \frac{nF}{RT}(E - E)^0,$$

and $\Gamma_T (= \Gamma_O + \Gamma_R)$ correspond to the total surface covered with reduced and oxidized sites. The other parameters have their usual meanings. Apart from the mirror symmetry of the waves, it is also characteristic that, in contrast to measurements obtained with dissolved redox systems, current i and the scan rate v are directly proportional to each other. However, if the heterogeneous kinetics are sluggish the two waves shift in relation to each other and, depending on the level of the k_s-values, become increasingly asymmetrical [138-140].

The above statements are valid for monomolecular layers only. In the case of polymer films with layer thickness into the μ-range, as are usually produced by electropolymerization, account must also be taken of the fact that the charge transport is dependent on both the electron exchange reactions between neighbouring oxidized and reduced sites and the flux of counterions in keeping with the principle of electroneutrality [141-144]. Although the molecular mechanisms of these processes are not yet understood in all detail, their phenomenology can be adequately described by the laws of diffusion [143]. All variants of the diffusive mass transport are possible, from the finite limiting case through to semiinfinite diffusion, depending on film thickness, the values of the formal diffusion coefficients, and the experimental time

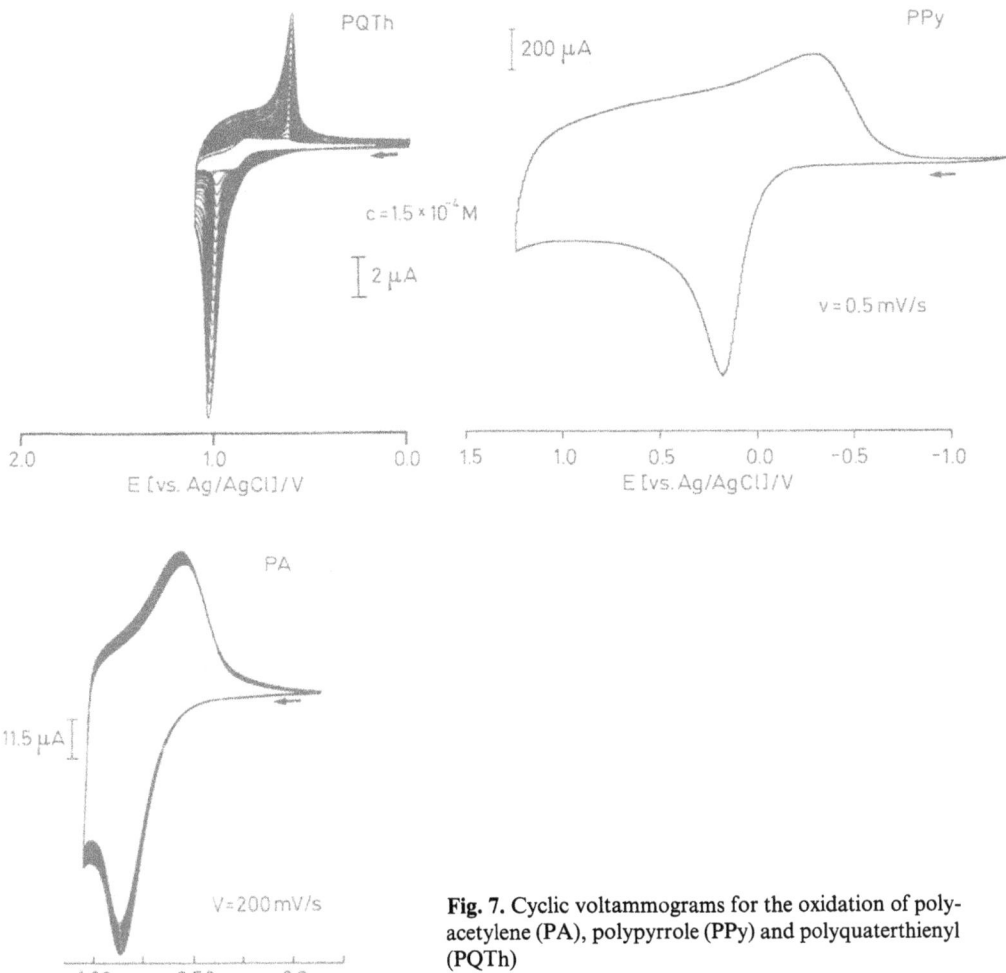

Fig. 7. Cyclic voltammograms for the oxidation of polyacetylene (PA), polypyrrole (PPy) and polyquaterthienyl (PQTh)

scales used. For voltammetric experiments this implies that as the sweep rate increases there must be a shift from mirror symmetrical CV diagrams with i proportional to v to the classical, asymmetrical voltammogram with i proportional to $v^{1/2}$.

Although the potentiodynamic charging and discharging of conducting polymers produces voltammograms of very different shape depending on type and polymerization conditions, one frequently finds CV diagrams (Fig. 7) with a very similar 'structure'. This is particularly the case when mechanically stable, free-standing films are produced from unsubstituted monomers under mild conditions [42,52,64,66,69,84,145]. Characteristic features of these systems are, in the case of the p-doping, a steep anodic wave at the start of charging followed by a broad, flat plateau as potential increases. In the reverse scan a potential-shifted cathodic wave appears at the negative end of the capacity-like plateau, whose peak current corresponds to approximately half the value of the anodic peak.

The conspicuous separation between the cathodic and anodic peak potentials was initially interpreted in terms of the simple theory for redox polymers as a kinetic effect of slow heterogeneous charge transfer; the thermodynamic redox potential of the whole systems was calculated from the mean value between E_{pc} and E_{pa} [52,64,99,146]. As a simplification, it was assumed that the interactions between charged oligomeric segments were negligible. These redox data correlate well with potential values obtained by extrapolation from quantum mechanical calculations [147] and redox potential measurements [67,148,149] on oligomers of defined chain length. As expected, with increasing conjugation length the respective oxidation and reduction potentials shift towards low energy values. However, the small discrepancy between the measured

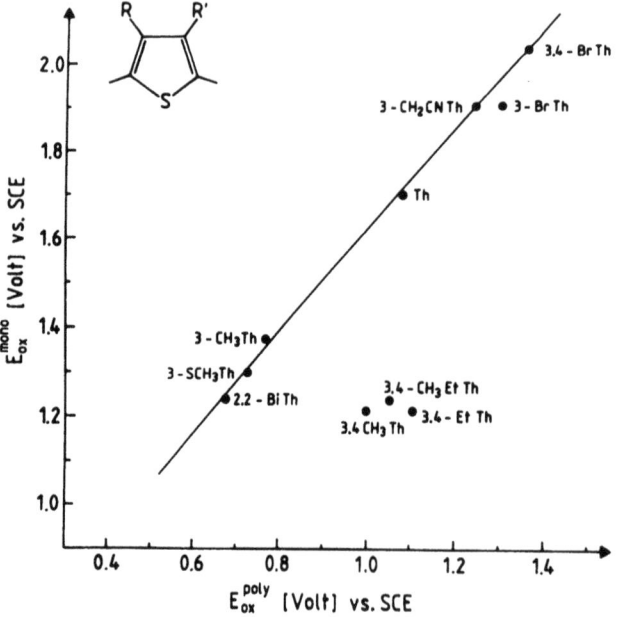

Fig. 8. Relationship between oxidation potential values of substituted monomeric thiophenes (E_{ox}^{mono}) and their corresponding polymers (E_{ox}^{poly})

and the extrapolated E_{pa} values gave rise to the opinion that the conjugation length of the neutral polymer covered only 5 to 10 monomeric units [148]. Furthermore, from the quantum mechanical findings it was concluded that that the onset potentials measured in the polymers reflected only the initial ionization of such systems and that further ionization steps had to be linked to structural changes in the system [147]. Substitution effects, too, influence the position of the oxidation potentials of the polymers. Whereas electron-donating substituents lower oxidation potentials just as they do in the monomers, the electron-withdrawing substituents raise them. The fact that in the majority of cases there is a linear relation between the oxidation potentials for the monomers and the polymers shows that the substituents' effects are primarily of an electronic nature [66,100,146,150] (Fig. 8).

Another concept has been developed on a refined model based on two-step redox systems typical for organic compounds [151]. This concept [152,153] treats a polymer chain with the degree of polymerization n as x weakly interacting segments containing k monomeric units, each of which can be charged up to a diionic state in two redox steps with separate potentials ($1 < k \ll n$, $x * k = n$):

$$M_n \underset{E_1^0}{\overset{+xe}{\rightleftarrows}} (M_k^\theta)_x, \tag{5a}$$

$$(M_k^\theta)_x \underset{E_2^0}{\overset{+xe}{\rightleftarrows}} (M_k^{2\theta})_x. \tag{5b}$$

This concept implies low interaction energies between the segments; consequently, the essential determinant of the position of the redox potentials must be the structure of the so-called effectively conjugated segments (ECS).

In principle, such propositions resemble the bipolaron model, which presents the physicist's view of the electronic properties of doped conducting polymers [153-159]. The model was originally constructed to characterize defects in solids. In chemical terminology, bipolarons are equivalent to diionic spinfree states of a system (S = 0)

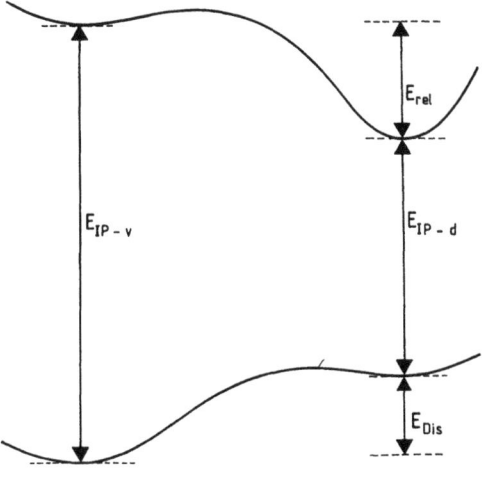

Fig. 9. Sketch of potential energy curves of a segment of conducting polymers in the ground state and in the ionized state; E_{IP-v} is the vertical ionization energy, E_{rel} the relaxation energy gained in the ionized state, E_{IP-d} the ionization energy of the distorted molecule, and E_{dis} the geometrical distortion energy in the ground state

after oxidation or reduction from the neutral state. The transition from the neutral state to the bipolaron state is via the polaron state (=monoion, S = 1/2, ESR signal) and, thus, corresponds sequentially as well to redox transitions observed in two-step redox systems. In contrast to normal redox processes, however, additional local distortions occur in the chain during charging of the polymer. Already in the first step, the formation of polarons, there is a gain of relaxation energy E_{rel}. This is released by structure relaxation after ionization, which has required a vertical Frank-Condon-like ionization energy E_{IP-v}. E_{rel} corresponds to the bonding energy of the polaron (Fig. 9).

The structural relaxation causes a local distortion of the chain in the vicinity of the charge, whereby the twisted benzoid-like structure of the affected segment transforms to a chinoid-like structure in which the single bonds between the monomeric units shorten and assume double-bonding character. Quantum mechanical calculations predict that polypyrrole in a polaron state of this nature must contain four pyrrole rings. Conversely, in discharging, the energy E_{IP-d} is first released; a coplanar ground state emerges, which then relaxes into the equilibrium geometry as the distortion energy E_{dis} is released.

Removing a second electron from the polymer segment results not in two polarons but in the bipolaron, which is predicted to be energetically more favoured than the polaron. The reason for this lies in the respective structural relaxations: that for the bipolaron is considerably greater than that for the polaron. The ionization is considerably greater than that for the polaron. The ionization energy required to remove a second electron decreases, and the electron affinity for taking up a second electron increases. The energy gain of the bipolaron compared to two polarons is said to be about 0.45 eV in the case of polypyrrole [160] and 0.35 eV in the case of polyparaphenylene [155] (Fig. 10).

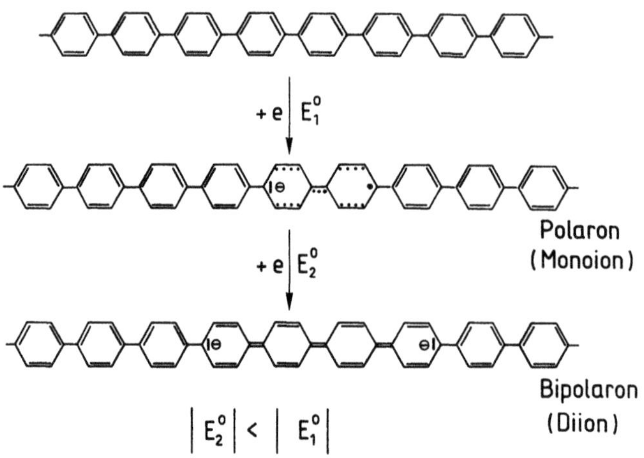

Fig. 10. Formation of the bipolaron (= diion) state in poly-p-phenylene upon reduction: In the model it is assumed that the ionized states are stabilized by a local geometric distortion from a benzoid-like to a chinoid-like structure. Hereby one bipolaron should thermodynamically become more stable than two polarons despite the coulomb repulsion between two similar charges

Whereas the intermediate existence of polarons has been unequivocally proved by ESR measurements and optical absorption data, up to now, the existence of bipolarons has been only indirectly deduced from the absence of the ESR signal and the disappearance of the visible polaron bands from the optical absorption spectrum [16-163]. On the other hand, spinfree — diionic-charge — states in aromatics, whose optical properties bear a remarkably resemblence to the predictions of the bipolaron model, have long been known [164-166]. Further evidence of bipolarons is the fact that doped polymers have a higher conductivity, which, even though not in complete agreement with the classical band model, can only be explained in terms of spinfree charge carriers [154,155,167,168].

So far, electrochemical measurements have not provided any direct proof for the formation of a bipolaron state in oligomers or polymers which is significantly more stable than the polaron state. In general, in terms of energy the redox potentials E_2^0 for bipolaron formation should be much lower than the potentials E_1 for polaron formation ($/E_2^0/ < /E_1^0/$). However, more recent electrochemical and ESR spectroscopic studies by Nechtschein et al. indicate that the bipolaron state is not much more stable than the polaron state [169,170].

Closely connected with the problems of the charge storage mechanism is the question to what extent one must and can distinguish between faradaic and capacitive processes during charging and discharging [64,171-173]. Feldberg [173], in particular, holds that it is hardly possible to draw such a distinction, and that the capacitive charge portion is proportional to the amount of the oxidized polymer. He concludes from the comparison with optical data that the faradaic oxidation processes in CV experiments are essentially limited to the range of the 'peak-shaped' anodic wave, whereas the adjoining broad anodic tail represents the large double-layer capacity of the system (Fig. 7). Independent impedance measurements by Bard et al. [171] and

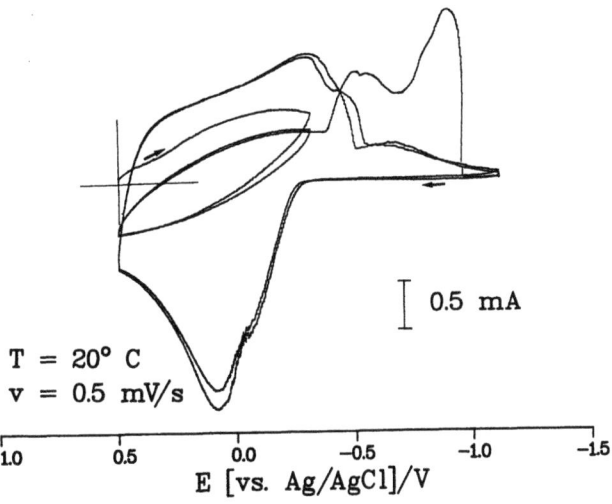

Fig. 11. Cyclic voltammetry of "first" discharging/charging of galvanostatically prepared PPy films (PC, 0.5 M LiClO$_4$); the first three cycles between +0.5 V and −0.3 V, the following between +0.5 V and −1.1 V

Barendrecht et al. [172] show that the charged polymer films are similar to porous metal electrodes with large double layer capacities.

In a refined model, Tanguy et al. [174] propose 'the existence of two ionic trapping sites'. One of these types is responsible for the deep trapping of the counterions which cannot follow the ac signal measurement and are released only at low potential in the reduction process. The other type of trapping is a shallow one in which the ions can follow the low frequency signal giving the capacitance effect. This capacitance is explained on terms of an ionic double layer formation at the surface of the polymer chain arising from the accumulation near this chain of the weakly trapped ions.

Voltammetric measurements on newly formed PPy films present a different picture. Apparently, during the first voltammetric discharging scan (v = 0.5 mV/s) only small currents are measured. The same effect also appears during the subsequent charging scan and can be generated in further cycles provided that the film is never totally discharged and retains at least 80% of it originally produced charge (Fig. 11).

When the switching potential for the discharging step is shifted to more negative potentials a pronounced cathodic wave in the range between −0.3 and −1.0 V vs. Ag/AgCl is observed; this wave subsides considerably during the following cycles. In the subsequent charging/discharging cycles the well-known voltammograms with their broad anodic tail appear. This unusual discharging phenomenon has been described several times [175−177].

On the basis of experimental findings Heinze et al. [175] propose the formation of a particularly stable, previously unknown tertiary structure between the charged chain segments and the solvated counterions in the polymer during galvanostatic or potentiostatic polymerization. During the discharging scan this structure is irreversibly altered. The absence of typical capacitive currents for the oxidized polymer film leads them to surmise that the postulated double layer effects are considerably smaller than previously assumed and that the broad current plateau is caused at least in part by faradaic redox processes.

Table 2. "Doping" levels for conducting polymers

Polymer	Counterion	Degree of "doping"	Refs.
PPy	ClO_4^-	0.30–0.33	65, 124, 178)
PPy	BF_4^-; PF_6^-	0.25–0.32	179)
PPy	$CF_3SO_3^-$	0.30	124, 179)
PTh	BF_4^-; PF_6^-	0.06	42, 49, 66, 91, 180)
PTh	ClO_4^-; BF_4^-; $CF_3SO_3^-$	0.30	124)
PTh	ClO_4^-	0.20–0.26	181, 182)
PBiTh	SO_4^{2-}	0.22	49, 66)
P-3MeTh	ClO_4^-; $CF_3SO_3^-$	0.25–0.30	124)
P-3MeTh	PF_6^-	0.12	49, 91)
PAz	ClO_4^-	0.25–0.28	42, 49, 105)
PFu	BF_4^-	0.26	42)
PPP	AsF_6^-; BF_4^-; PF_6^-	0.16	183–185)
PPP	PF_6^- (SO_2)	0.27	94)
PPP	Li^+	0.44	184, 185)
PANI	Cl^-	0.42	186–188)
PA	BF_4^-; PF_6^-; ClO_4^-	0.06–0.08	84, 189–191)
Pa	Li^+; Na^+; K^+	0.07–0.18	189, 191)

A further important characteristic for charge storage in conducting polymers is the amount of doping or insertion. This gives the mole fraction of the corresponding monomers whose charge is compensated by incorported counterions. It is determined either by elementary analysis or by coulometric measurements. However, these measuring techniques may produce different results as the elementary analysis cannot distinguish an additional solvent portion, and coulometry also includes the capacitive charging. The optimum doping level for PPy is about 0.33, but can have very much lower values. This depends, inter alia, on structure and the applied charging potential, but is also influenced by environmental parameters such as the solvent or the supporting electrolyte (Table 2). Spectroelectrochemical studies [192], electronic quartz microbalance (EQCM) measurements [193] as well as SIMS measurements [194] on the mechanism of ion transport during charging and discharging of conducting polymers prove that p-doping involves not only the incorporation of anions but also cations, the latter largely during reduction of the film. The authors propose a two-step mechanism. During the first discharging step, cations are incorporated into the polymer, forming ion pairs with the mobile anions; in the second step surplus anions as well as the ion pairs slowly diffuse out of the film:

$$[(PPy)_4^+ A^-]_n + (n-m)\,e^- + (n-m)\,M^+ \rightarrow$$
$$[(PPy)_{4n}^{m+} mA^-] + (n-m)\,[M^+, A^-]_{pol}, \tag{6a}$$

$$[(PPy)_{4n}^{m+} mA^-] + me^- \rightarrow (PPy)_{4n} + mA^-_{sol}$$
$$(n-m)\,[M^+, A^-]_{pol} \rightarrow (n-m)\,[M^+, A^-]_{sol}. \tag{6b}$$

In a recently published study Miller et al. [195] describe similar phenomena in which XPS and AA measurements prove the incorporation of cations during the discharging of PPy.

From the respective doping levels one can deduce that in the oxidation of PPy, and often of PTh as well, every 3rd or 4th heterocycle is charged, whereas in the case of PPP, provided that the experiments take place in common solvents such as propylene carbonate, only every 6th monomeric unit is charged. By contrast, if one uses SO_2, a solvent with low nucleophilicity, PPP can be reversibly oxidized to a doping level of 0.24, which corresponds formally to a charge on every 4th monomer unit [196]. The differences in chargeability of PPP and PPy are explained by the high oxidation potential of PPP. This significantly increases the reactivity of this system towards the solution and other nucleophilic impurities. One should note that polymers such as PA and PPP can be more highly charged by reduction (n-doping) than by oxidation. This illustrates the advantages of solvents such as propylene carbonate and THF, which stabilize the anions particularly well.

The above findings are in approximate agreement with the theoretical calculations for the bipolaron model. These predict that the bipolaron state (=diion) in PPy [160] extends over 4 pyrrole rings, and in PPP [155] over 5 rings.

Voltammetric measurements on defined soluble oligomers in the homologous series of the *p*-phenylenevinylenes and the *p*-phenylenes [196,197] have finally ended speculations about the nature of the redox processes in conducting polymers. In particular, studies on the oligo-*p*-phenylenevinylenes show that the number of possible redox steps rises with increasing chain length (Fig. 12). The unequivocal experimental data

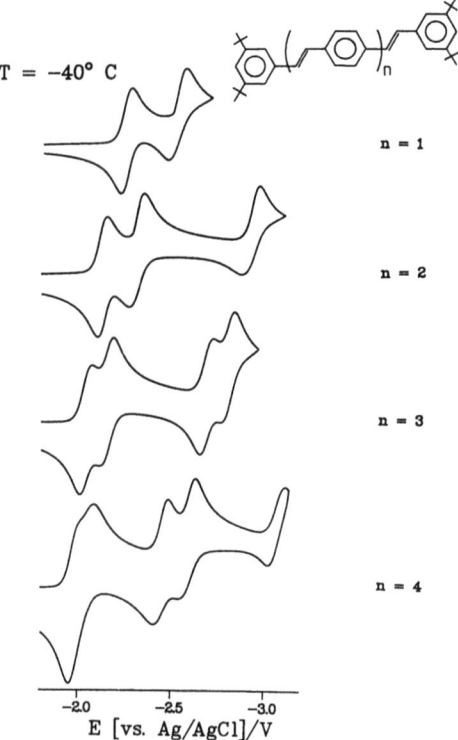

Fig. 12. Cyclic voltammograms (corrected with respect to background currents) for the reduction of defined oligo-*p*-phenylene-vinylenes (THF, NaBPh$_4$)

allow some general statements on the thermodynamics of the redox states in conducting polymers. First, with increasing polymer chain length the energies of different redox stages gradually shift towards a uniform, low, limiting potential. Second, the redox states degenerate in pairs with increasing chain length. Thirdly, adding successive monomeric subunits in the molecular chain enlarges the number of accessible redox states in agreement with expectations. However, the energetic gap increases strongly between the lowest and the highest charged states.

For real polymer systems — characterized by a more or less broad distribution of molecular weight and, depending on the conditions of formation, various structural defects — this means that at the start of charging there is a high density of virtually degenerate redox states. As potential increases the density of the redox transitions decreases but remains finite within the experimentally accessible potential range. This explains the broad, plateau-like waves which are so often characteristic of the potential range following the peak-like main wave in voltammetric experiments. Digital simulations carried out under the assumption of mutual overlapping of such redox processes support this interpretation [197].

Insofar as the additional stabilization of the diionic state predicted as a consequence of a geometric distortion is not observed, the experimental data for oligo-*p*-phenylenevinylene and oligo-*p*-phenylenes are not fully consistent with the theoretical predictions of the bipolaron model. Nevertheless, the charging and discharging of conducting polymers obviously does involve changes in geometry. The best documented of these is the *cis/trans*-isomerization during the electrochemical oxidation of *cis*-PA, for which there are IR, Raman, ESR and UV measurements [198-201]. By contrast, it is very difficult to spectroscopically identify the geometric distortions of polymer chains and the formation of chinoid structures in systems such as PPy and PTh. But reports of electrochemical experiments [202,203] are sufficient proof of redox processes causing structural changes to provide a basis of comparison adequate for deducing analogous processes in conducting polymers. In agreement with this knowledge of conformational changes during electrode reactions of monomeric species, Heinze et al. [197] interpret anodic current-voltage curves of PPy as charging processes in which a geometric distortion of segments of polymer chains is accompanied by the formation of chinoid-like structures already at the monoionic level (polaron state). Under these preconditions, the energy needed in further electron-transfer reactions to produce diionic states (bipolaron) is virtually the same as that for the formation of the monoionic polarons. During discharging the coplanar chain structures initially remain unchanged. The system relaxes into a twisted equilibrium conformation only after complete reduction to the neutral state. In consequence, the discharging reduction potentials shift to negative values — the charged system is energetically stabilized — so that the resulting cyclic voltammograms as a whole are characteristically asymmetrical. The results of digital simulations based on this

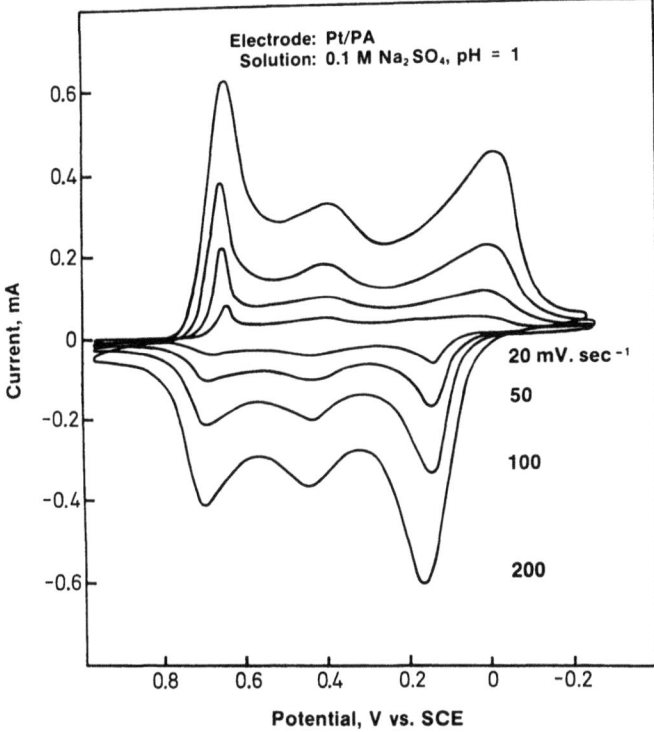

Fig. 13. Cyclic voltammogram for the charging/discharging of polyaniline on Pt substrate at different scan rates, ref. 214 (reprinted by permission of the publisher, The Electrochemical Society, Inc.)

charging/discharging model agree closely with those of the cyclic voltammetric experiments.

The information from electrochemical voltage spectroscopy (EVS) on thermodynamic parameters [204-206] is in essence the same as that obtained by cyclic voltammetry. This technique involves slowly stepwise incrementing the voltage of the electrochemical cell and recording the charge removed from (or injected to) the polymer after each voltage step. Since the cell current in EVS measurements can be kept arbitrarily low, non-equilibrium effects such as IR-losses, over-potentials, and diffusional phenomena are virtually eliminated. For the conducting hetero-aromatic polymers a typical hysteresis in charging and discharging cycles using the EVS technique is observed [207-209]. This intrinsic hysteresis corresponds to the asymmetry between the oxidation and reduction potentials in the CV experiment. It can be evaluated as additional evidence of the geometric relaxation effects during charging and discharging of conducting polymers.

Electrochemical measurements on polyaniline (PANI) produce a picture of the charge storage mechanism of conducting polymers which differs fundamentally from that obtained using PTh or PPy. In the cyclic voltammetric experiment one observes at least two reversible waves in the potential range between -0.2 and $+1.23$ V vs SCE. Above $+1.0$ V the charging current tends to zero. 'Capacitive' currents and over-oxidation effects, as with PPy and PTh, do not occur [210-212] (Fig. 13). The striking changes in the charge characteristics of polyanilin compared with PPy and PTh are largely explained by the nitrogen atoms capturing a considerable part of the positive charge, which eliminates most of the coulomb interaction between the charged centres.

PANI is usually produced by the anodic oxidation of aniline in acidic aqueous solution [52,213-217], but can also be produced by chemical oxidation [211,218]. Hence, it is not surprising that the oxidation of PANI is pH-dependent, and that therefore, in addition to the electron-transfer processes, proton-transfer reactions occur during charging. Although it is usually assumed that PANI has a chain structure (emeraldine) with head-tail connection between the aniline units [221], the existence of cyclic structures has also been postulated in the literature [219]. It is generally accepted that there are different, possibly coexisting forms of polyaniline, including benzoid and chinoid rings, free amines (NH), imines (=N) and protonic amines and imines [220]. At present, discussion centres on three different models of the charge storage mechanism in polyaniline [211,222,223]. Depending on the pH-value of the solution, differently protonated structures are formed. In the most simple case [222], which is supported by quantum mechanical calculations, it is assumed that poly-radical cationic states (polarons) are formed in the first step.

In the second step, which correlates with the second oxidation peak in the CV diagram, four protons and two electrons are transferred.

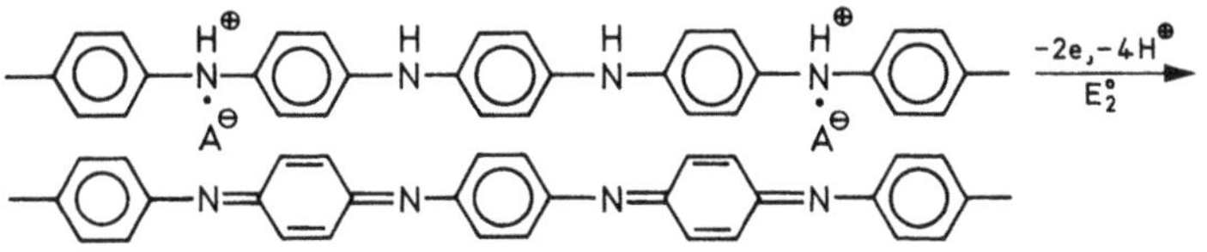

[Scheme: oxidation of partially protonated PANI structure, $-2e, -4H^\oplus$, E_2°]

This corresponds with MacDiarmid's observations [211], which show that the second redox step is strongly pH-dependent. MacDiarmid further differentiated his redox model to take account of the fact that pure leucoemaraldine with its amine-N is already protonated at pH values ≥ 2, and that the totally oxidized pernigraniline with its less basic imine-N can also be protonated. This gives the following (simplified) reaction scheme:

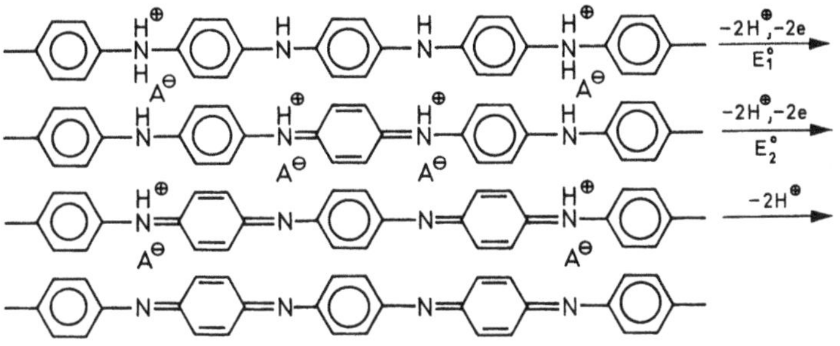

[Scheme: three successive oxidation steps with $-2H^\oplus, -2e$, E_1°; $-2H^\oplus, -2e$, E_2°; $-2H^\oplus$]

The third variant is that of Genies and Lapowski [223]. They base their model on the assumption of fundamental differences between the two oxidation steps in PANI. The first step is analogous to that of MacDiarmid. For the second redox process the authors postulate the additional oxidation of imine-N

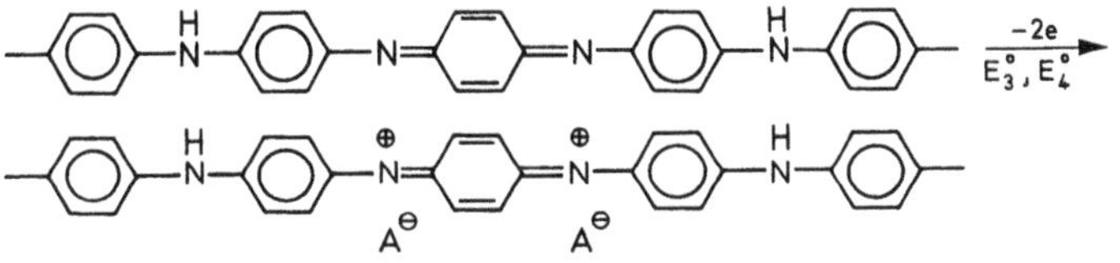

[Scheme: $-2e$, E_3°, E_4°]

This approach finds experimental support in FTIR measurements of the oxidation of PANI in organic solvents [224], which indicate an anion intercalation mechanism for the second oxidation step. However, the IR findings may also be interpreted as support for the formation of a protonated imine structure [224].

5 Applications

The enormous efforts put into the basic research and development of conducting polymers are naturally related to hopes of feasible technical applications [225-227]. The starting point of this development was the discovery that PA can function as an active electrode [9] in a rechargeable polymer battery. Since then, the prospects of technical application have grown considerably [228]. Apart from the battery electrode, conducting polymers are discussed as potential electrochromic displays [229], information memories [225a,230], antistatic materials [23,231,232], anti-corrosives [10,233], electrocatalyzers [234], and as materials in molecular electronics [225b,228] and biomedicine. The most advanced developments are the battery [190,191] and electrochromic displays [229].

5.1 Rechargeable Batteries

The development of a rechargeable polymer battery is being pursued worldwide. Its attraction lies in the specific weight of polymers, which is considerably lower than that of ordinary inorganic materials, as well as potential environmental benefits. In principle there are three different types of battery. The active polymer electrode can be used either as cathode (cell types 1, 2), or as anode (cell type 3), or as both cathode and anode (cell type 4) (Fig. 14). As the most common polymer materials are usually only oxidizable, recent research has concentrated on developing cells with a polymer cathode and a metal anode.

The polyacetylene cell has been most extensively researched [9,189,235-243]. Initially, p-doped (oxidized) PA was used as active electrode, and Li as anode [9,189,238,241], in cells containing inter alia solutions of $LiClO_4$ in propylene carbonate (PC) [9,189,238,241] or sulfolane [239] (cell type 1). Before PA cells can be used as current sources they first have to be charged. If $(CH)_x$ is the anode and Li^+/Li is the cathode, the overall charging reaction is:

$$(CH)_x + (xy)Li^+ \rightarrow (CH^{y+})_x + (xy) Li \tag{7}$$

The observed open circuit potentials (V_{oc}) lie between 3 and 4 V vs. Li^+/Li for a maximum charging of 6 mol%. The energy densities of such cells reach values of up

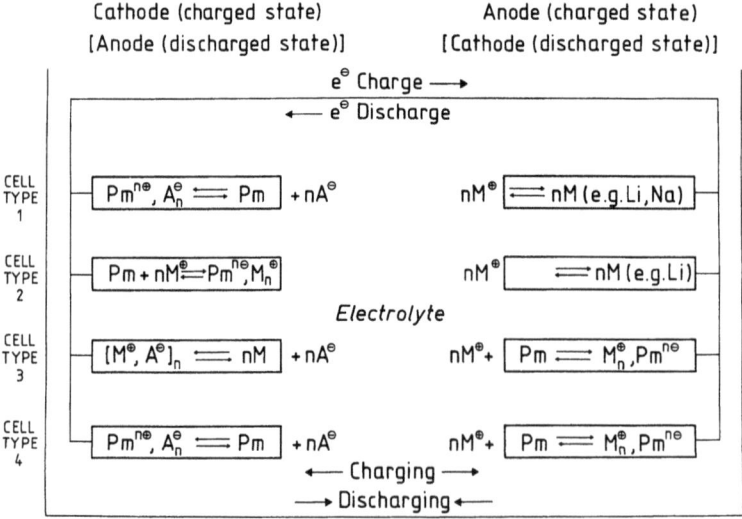

Fig. 14. Cell types for a polymer battery with active polymer anode, cathode or both respectively

to 120 Whkg^{-1}, which are far above the figures for conventional lead batteries (40 W h kg^{-1})[249]. Current densities can rise to 20 mA cm^{-2} during discharging; but in this event discharge capacity recovery is only 16%. At low current densities, up to 65% of total capacity can be recovered, which is about 85 A h kg^{-1} for a 6 mol% charging[189]. In an experiment to improve cycle stability a Li—Al alloy was substituted for the Li-anode of the conventional battery[249]. A solid state battery employing oxidized PA as the cathode-active material has also been reported[243]. The cell, Ag/RbAg$_4$I$_5$/(CH^{y+}) (I$^-$)$_{yx}$ operates at room temperature with a V_{oc} of 0.65 V, an energy density of 10 W h kg^{-1}, a current density up to 50 mA cm^{-2} of (CH)$_x$ and a long lifetime, with a test cell functioning for at least two years.

As the reduction potential for (CH)$_x$/(CH^{y-})$_x$ is approx. 0.5 to 1 V positive to the Li$^+$/Li electrode, reducing neutral PA with a positive EMF produces an energy gain (cell type 2)[244-246,250,251]. The complete cell reaction is then:

$$(CH)_x + (xy)\,Li + (Bu_4N^+)(ClO_4^-) \rightarrow [(Bu_4N^+)_y(CH^{y-})_x + (xy)\,LiClO_4. \tag{8}$$

However, under these conditions, the energy density is lower than with the (CH^{y+})$_x$ electrode, whereas, on the other hand, the coulombic efficiencies attain values of 98%. In some cases PA in its n-doped form has also been used as battery anode (cell type 3)[190,191,241]. The discharging reaction in a cell with a TiS$_2$ cathode is as follows:

$$[Li_y^+(CH^{y-})] + (TiS_2)_x \rightarrow [Li_y^+(TiS_2^{y-})]_x + (CH)_x \tag{9}$$

This cell has an open circuit voltage of 1.65 V and possesses good stability. It has a theoretical energy density of 110 W h kg^{-1}.

There have been several reports of all-plastic batteries with PA-electrodes (cell type 4)[9,249,252,253]. The observed cell potentials lie between 3.4 and 2.5 V, the short circuit current was 50 mA cm^{-2} down to 12 mA cm^{-2}. The overall discharge reaction is:

$$(CH^{y+})_x + (CH^{y-})_x \rightarrow 2(CH)_x \tag{10}$$

Although, in principle, the properties of PA are promising, its commercial prospects are less so. This is chiefly due to the chemical instability of PA, which is increased by doping. Highly conducting PA with conductivities of over 100000 S/cm[254] has recently been developed, but this has not improved the prospects of polymer batteries. Among the essential requirements are charging capacity, cycle stability and the self-discharging rate. Studies have shown that doped PA in vacuum spontaneously decomposes at temperatures of 60 to 80 °C, forming BF$_3$ and PF$_5$ from the anions BF$_4^\ominus$ and PF$_6^\ominus$[188]. In PC, the most extensively used solvent in batteries, the material has largely "degenerated" after just 10 charging-discharging cycles[255]. Independent experiments with the largely inactive solvent SO$_2$ prove[84] that the nucleophilic reactivity of the solvent[248] is a basic cause of PA degrading within a few battery cycles, as it leads to addition reactions, thereby interrupting conjugation (Fig. 15).

A promising candidate for a polymer battery that does not possess the typical disadvantages of PA is PPy[176,178,256-259]. Its open circuit voltage lies near 3.5 V vs Li$^+$/Li. Charge capacity is about 70 to 85 A h kg^{-1}, and the effective energy density

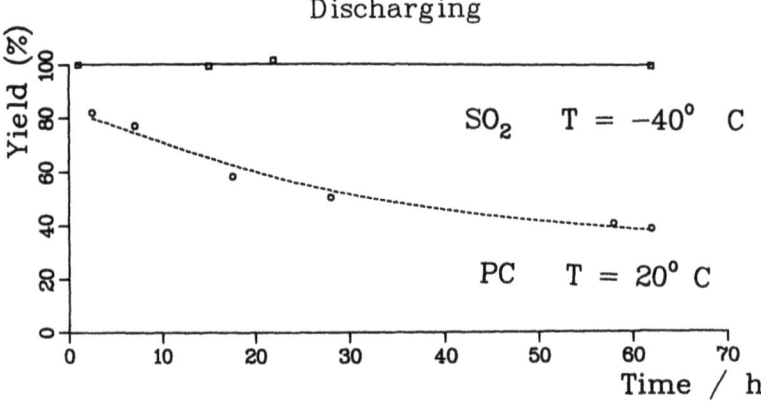

Fig. 15. Self-discharge of oxidized PA in SO_2 (charged to 5.5 mol%) and in PC (charged to 4.5 mol%)

(vs Li^+/Li) reaches values of 40 to 60 W h kg^{-1}. In addition, PPy has a long lifetime and a relatively slow self-discharging rate. PTh has similar properties to PPy [182, 260-262]. Its V_{oc} is +3.5 V vs Li^+/Li and its charge capacity lies near 70 A h kg^{-1}. However, its self-discharging rate of 10 to 15%/d is extremely high. Poly(p-phenylene) (PPP) has also been studied as a potential battery electrode [94,183,184]. As the neutral form can be both oxidized and reduced, it can in principle serve both as a positive and a negative electrode. The p-doped material gives an open circuit voltage of 4.4 V vs a Li-electrode and a current density of 40 mA cm^{-2}. With an electrochemically produced film the charge capacity reaches values of 90 A h kg^{-1} [94]. The powder obtained by chemically oxidizing benzene is not very conductive and must first be predoped [183].

Other materials which appear to be suitable as active battery electrodes are polyazulene [263a], poly (N-vinylcarbazole) [263b], polyquinolines [264] and, most recently, polyaniline (PANI) [265-268]. PANI, in particular, is a promising material for the construction of a commercial battery. It can be used with both aqueous and aprotic electrolytes and has a high coulombic efficiency with high cycle stability. In aqueous electrolytes PANI can be employed both as cathode (cell type 1) and as anode (cell type 3):

PANI/$ZnSO_4(H_2O)$/Zn ,
PbO_2/$ZnSO_4(H_2O)$/PANI .

The first cell has the maximum capacity of 108 A h kg^{-1} and the energy density of 111 W h kg^{-1}. The coulombic efficiency was close to 100% over at least 2000 complete cycles when cycled between 1.35 V and 0.5 V at a constant current density of 1 mA cm^{-2}. The second cell also showed excellent recyclability (4000 cycles with 95% coulombic efficiency), on the other hand the discharge capacity decreased steadily from 40 to 25 A h kg^{-1} after 4000 cycles. In PANI batteries with aprotic electrolytes PC is usually employed as solvent and Li as the negative electrode [210,266-268]:

PANI/PC($LiClO_4$)/Li:

The open circuit voltage lies near 3.40 V with a discharging capacity of 140 A h kg^{-1}. It is stated to have an energy density of 352 W h kg^{-1} and to show excellent rechargeability [210].

An important problem encountered with polymer electrodes is that of overoxidation. It occurs after reversible charging of the electrode at high oxidation potentials and leads to polymer degeneration. The results of thorough studies [84,248] show that such degenerative mechanisms are promoted by the nucleophilicity of the solvent. Especially the activity of water leads to the formation of quinone-type compounds, to the cleavage of C—C bonds, the liberation of CO_2, and the formation of carboxylic acids [269-272]. Hence, there is a clear tendency to avoid both nucleophile solvents and water in the construction of polymer batteries.

5.2 Electrochromic Devices

Apart from polymer batteries, applied research on conducting polymers is most advanced in the field of electrochromic displays [225,229,273]. Electrochromic displays are based on an electrochemical reaction of a material that displays a visual change upon changing its redox state. The advantages of using electrochromic displays (ECD) are their ease of preparation and the uniformity of the prepared films. Furthermore, ECDs have no limited visual angle but they exhibit a memory function, even after the driving voltage has been removed. Erasing and rewriting can be performed almost at will.

The suitability of PPy [73,274,275], PTh [145,276] and PANI [277-279] as electrochromic displays have been intensively studied. Recently, the suitability of other polymers [227] such as polyisothianaphthalene [55,280] has also been analyzed.

PANI appears to be particularly promising, as its colour changes during oxidation covers the spectrum from yellow through green, blue and violet to brown. The response times for the transition from one state to the next lie below 100 msec, which is acceptable for practical applications. The maximum cycle count so far is about 10^5, and thus one order of magnitude worse than that for inorganic tungsten oxide. Whether such devices can be technically realized depends on whether the life times of polymers can be lengthened and whether response times can be shortened by a selection of suitable anions which diffuse into the polymer.

5.3 Miscellaneous

The unusual electric and optical properties of conducting polymers together with the typical mechanical material properties of polymers have raised expectation of numerous further applications in the future. As the mechanical stabilities of conducting polymers such as PPy or PANI are generally poor, it is hoped to overcome this by synthesizing composites. Two techniques have been developed. In one case, pyrrole is electropolymerized in a non-conducting polymer film such as PVC, which is prepared on the electrode by dipcoating [232,281-285]. The resulting composites have the desirable mechanical properties of the PVC host, yet they retain the high conductivity of the PPy dopant. In the same way, films have been produced which contain poly(vinylalcohol) [286] or polyether/polyester [287] as insulating polymer materials.

Stable highly conductive films of PPy-Nafion-impregnated Gore-tex [288] and PPy-Nafion [289,290] have also been reported. Polymer blends can also be produced by coelectropolymerizing suitable host compounds. This composite technique includes the electrooxidation of pyrrole with pyrrole-grafted polymers [291], the electropolymerization of α,α'-bithienyl in THF [292] and the anodic oxidation of 3-methylthiophene in a solution containing poly(methacrylate) [293]. A chemical variant for producing composite materials consists of dispersing the conducting polymer in the host matrix. PANI/PVC blends are produced in this way [294]. Potential applications discussed in the literature include antistatic materials, electromagnetic shieldings, or reversible fuses [28,231,295].

Intensive research on the electrocatalytic properties of polymer-modified electrodes has been going on for many years [13,14]. Until recently, most known coatings were redox polymers. Combining redox polymers with conducting polymers should, in principle, further improve the electrocatalytic activity of such systems, as the conducting polymers are, in addition, electron carriers and reservoirs. One possibility of intercalating electroactive redox centres in the conducting polymer is to incorporate redoxactive anions — which act as dopants — into the polymer. Most research has been done on PPy, doped with inter alia Co^{2+} [296], RuO_4^- [297], Co- and Fe-phthalocyanines [298,299] as well as Co-porphyrines [300,301]. Evidently, in these cases the cathodic activity for the O_2-reduction is significantly increased. A further method of incorporating redoxactive groups in PPy is to covalently bind the redox centre to the pyrrole nitrogen via an alkyl chain [302-310].

Increasing attention is being paid to research on conducting polymers as materials for microelectronic devices. Here the characteristic property — changing conductivity by up to 9 orders of magnitude through doping — is exploited for the construction of diodes and transistors. The derivatized electronics were fabricated e.g. on an array of Au microelectrodes by depositing polypyrrole [315,316], poly(3-methylthiophene) [12] and polyaniline [317]. The polymer-based transistor-like devices can be turned on and off by electrical or chemical signals that oxidize or reduce the polymer. The charge necessary to turn the device completely on is 10^{-6} C. The devices have transconductance values of only about 1 order of magnitude lower than found for good solid-state Si MOSFET's. They are stable for thousands of on-off cycles and have switching times below 50 ms.

In recent years further concepts have been developed for the construction of polymer-based diodes, requiring either two conjugated polymers (PA and poly(N-methylpyrrole) [318] or poly(N-methylpyrrole in a p-type silicon wafer solid-state field-effect transistor [319]. By modifying the transistor switching, these electronic devices can also be employed as pH-sensitive chemical sensors [320], or as hydrogen or oxygen sensors [321] in aqueous solutions. Recently a PPy alcohol sensor has also been reported [322].

Electron transfer processes at the electrode/electrolyte interface take place not only when there is a suitable electrode potential, but can also be activated by photoenergetic processes. These photophsysical or photovoltaic processes occur mainly in semiconductors or semiconductor-like materials. They convert light energy into electricity or chemical energy. The development of photoenergy conversion systems called solar cells is limited mainly by the problem of photocorrosion of small bandgap electrodes. This degradation can be slowed or prevented completely by suitable coatings on the electrode. The highly conductible polymers like PPy [323-332] and

Table 3. Prospective applications of conducting polymers

Application	Materials	Refs.
Rechargeable Battery	Doped PA — Li system	9, 189, 238, 241)
	Doped PPy — Li system	178, 256–259)
	Doped PANI — Li system	210, 266–268)
Polymer composites	PPy/PVC	232, 281–285, 295)
as antistatic and electromagnetically shielding materials	PANI/PVC	28, 294, 295)
Electrochromic device	PTh	145, 275, 276)
	PANI	277–279)
Electronic devices	Au/PPy	315, 316)
(diodes, transistors)	Au/P (3-MeTh)	12)
	Au/PANI	317)
	GaAs/P (3-MeTh)	335)
	Si/SiO$_2$/P (N-MePy)	319)
Chemical sensors: pH, O$_2$, H$_2$, alcohol	Pt/PANI; Pt/PPy	325–322)
Electrocatalysts	Modified PPy electrodes	296–310)
Piezoelectric transducer	Ceramics/PPy	336)
Memory devices	PPy	230)
Photoelectrochromic dev.	n-Si/P (N-MePy)	275, 337)
Photovoltaic devices	Si/PPy	323–329)
	CdS/PPy	330)
	GaAs/P (3-MeTh)	333, 335)
	PPy/liquid junction	338, 339)
	PANI/liquid junction	340)
	PTh/liquid junction	341)
Controlled drug release	P (MeOTh)	133)

PTh [333–335] are evidently materials which are suitable for such applications and which stabilize the photoanodes considerably.

Some of the most important applications for conducting polymers which might show at least some commercial viability in the near future are listed in Table 3. The list is by no means complete, and is growing all the time. However, one should not expect fundamental progress in practical applications until basic research on conducting polymers moves beyond the stage of trial and error, and develops concepts to obtain quantitative information about molecular structures and properties, on the one hand, and the resultant material properties on the other hand.

6 Acknowledgements

The authors wishes to thank the Deutsche Forschungsgemeinschaft, the Volkswagen Foundation, the Fonds der Chemischen Industrie and the Federal German Ministry of Research and Technology for their generous support of his research mentioned in this paper. This progress would have been impossible without the application, thoughts and perseverance of his coworkers, in particular Dr. John Mortensen, Dr. K. Hinkelmann, M. Dietrich and M. Störzbach. Finally, thanks are due to K.-O. Lorenz for his assistance in preparing the text.

7 Note Added in Proof

Since completion of the manuscript several special reviews have appeared on the electrochemistry of conducting polymers and related topics [342-344]. A further stimulus was the International Conference on Science and Technology of Synthetic Metals in Santa Fe 1988 (USA) [345]. Several trends are discernible. First, much effort is still concentrated on synthesizing novel heterocyclic monomers as well as subsituated derivatives of pyrrole and, especially, of thiophene in the hope of discovering interesting new material properties. Second, it is becoming apparent that optimization of polymerization conditions may considerably improve the qualities of polymers. Third, with respect to basic research, the combination of electrochemical techniques and spectroscopic methods has proved to be a powerful tool for characterizing the electronic and structural properties of these materials.

Several novel polymers such as poly[2,5-bis(2-thienyl)thiazole [346], poly(dihydrobenzodipyrrole) [347], poly-PyPy[3,2-b] [348], polydithienobenzene [349], polyisonaphthothiophene (PINT) [350], and polyisothianaphthene (PITN) [55] have been synthesized by electropolymerization. It was predicted [351] that some of these systems should, on electrochemical switching to the oxidized state, develop a very small band gap leading to transparent highly conducting materials. As was shown by Wudl et al. [344, 352], the band gap of charged PITN and PINT reduces to approximately 1.2 eV in the near infrared, which opens up interesting perspectives of applications as antistatic transparent films.

In the field of soluble conducting polymers new data have been published on poly(3-alkylthiophenes [353-355]. They show that the solubility of undoped polymers increases with increasing chain length of the substituent in the order n-butyl > ethyl ≫ methyl. But, on the other hand, it has turned out that in the doped state the electrochemically synthesized polymers cannot be dissolved in reasonable concentrations [356]. In a very recent paper Feldhues et al. [357] have reported that some poly(3-alkoxythiophenes) electropolymerized under special experimental conditions are completely soluble in dipolar aprotic solvents in both the undoped and doped states. The molecular weights were determined in the undoped state by a combination of gel-permeation chromatography (GPC), mass spectroscopy and UV/VIS spectroscopy. It was established that the usual chain length of soluble poly(3-methoxthythiophene) consists of six monomer units.

Meanwhile, the R-R coupling (see Sect. 2.2) has evidently found general acceptance as the main reaction path for the electropolymerization of conducting polymers [345]. The ionic character of the coupling species explains why polar additives such as anions or solvents with high permittivity [358] accelerate the rate of polymerization and function as catalysts. Thus, electropolymerization of pyrrole is catalyzed in CH_3CN by bromide ions [359] or in aqueous solution by 4,5-dihydro-1,3-benzenedisulfonic acid [360, 361]. The electrocatalytic influence of water has been known since the work of Diaz [69, 65] and is also described by other authors [76]. The effects of the extremely high dielectricity constant (180) of N-methylformamide (NMF) increase polymerization rates of PPy. Thus, in acetonitrile the relative growth rate of a PPy film in the presence of NMF increases by a factor of up to 300 [77b].

The discussion on the "capacitive" charging of conductive polymers continues [362-364]. Following Feldberg's hypothesis [173], the capacitive charge should be proportional to the amount of oxidizable film whereby it is assumed that oxidation of the film occurs at one defined redox potential E^0. All available experimental findings are unequivocal evidence [196, 197] that the latter assumption is wrong. In a recent paper Heinze et al. [365] again showed by spectroelectrochemical measurements that charging of PPy in the plateaulike range of current-voltage curves corresponds to the formation of bipolaron (iionic) states, which is equivalent to a faradaic charging process.

Simultaneous ESR and electrochemical measurements on a polypyrrole film give convincing evidence that the charging process in this film involves the generation of paramagnetic species which are obviously intermediates in the process of switching from the neutral to the oxidized state [366]. In any case, independent of all other findings, this proves that there are several redox states formed at different redox potentials. In a recent paper, Feldberg and Rubinstein [367] offer a new explanation for the typical finite difference between anodic and cathodic peak potentials of conducting polymers which is independent of scan rate in cyclic voltammetry. In their opinion the hysteretic behavior is not due to the classical "square scheme" involving heterogeneous and homogeneous kinetics, but to N-shaped free energy curves as a consequence of phase transitions in the polymer. The discussion on this topic is just starting in the literature, and a challenging exchange of views may be expected.

The cis/trans isomerization of cis-polyacetylene, previously only disclosed from spectroscopic data, has recently been detected by cyclic voltammetry [365]. The analysis of the redox data reveals that the trans-form is thermodynamically more favorable in the charged than in neutral state.

The electrodesposition process of conducting polymers can be monitored by spectroelectrochemical in situ techniques [344, 368-371]. Especially useful are ellipsometric studies [369] combined with time-resolved UV-visible spectroscopy [368, 371]. Measurements of the deposition of polythiophene reveal that during the starting period of electrolysis only species in the solution phase are generated and that the growth of the nuclei is not a simple three-dimensional deposition process but a fairly complex one.

The study of composite materials containing both conducting polymers and other polymeric systems attracts much interest because there are well-founded expectations of improving mechanical, electrical or other properties [372-375]. Obviously, the copolymerization of pyrrole and thiophene derivatives yields products which may be well suited for applications in photocorrosion and battery techniques [376, 378].

Research on the electrochemistry of polyaniline is steadily increasing, as documented by numerous publications [379-387]. Shacklette et al. [387] showed in a very thorough study of the phenyl-end-capped tetramer of polyaniline that the first oxidation step involves a 2e-transfer step leading to the emeraldine salt form in which 50% of the nitrogens are oxidized. The second oxidation wave also indicates a 2e-step and varies with pH at a rate of approximately 120 mV/pH, which suggests a deprotonation of 4 protons. In this case the resulting species should be the pure imine form.

The application of PANI as active electrode material in a commercially available polymer lithium battery is described by Nakajima and Kawagoe [388]. In aprotic solvents redox properties are optimal when PANI is switched from the amine to the emeraldine salt form and vice versa. Further constructions of rechargeable batteries using PANI as positive electrode are given by Mizumoto et al. [389] and Genies et al. [390]. Besides PANI, PPy [391, 392] is still under study for battery applications. Recently, poly(3-methylthiophene) was successfully tested as a rechargeable stable material [393, 394].

In a recent paper [395] it has been reported that unsaturated polymers such as polyisoprene which are not conjugated can be also reversibly doped by oxidation with I_2 or Br_2. In the doped state conductivities of more than 10^{-2} S/cm have been measured. Obviously, the belief that conjugated chains in polymers are a prerequisite for conduction can no longer hold. As yet there are no independent, electrochemical measurements which give further insights into this unusual phenomenon.

8 References

1. Shchegolev IF (1972) Phys. State. Solidi A 12: 9
2. Garito AF, Heeger AJ (1974) Acc. Chem. Res. 7: 232
3. Goodings EP (1976) Chem. Soc. Rev. 5: 95
4. Goddings EP (1975) Endeavour 73: 123
5. Perlstein H (1977) Angew. Chem. 89: 534; (1977) Angew. Chem. Ed. Int. Ed. Engl. 16: 519
6. Shirakawa H, Louis EJ, MacDiarmid AG, Chiang CK, Heeger AF: J. Chem. Soc., Chem. Commun. 1977: 578
7. Chiang CK, Park YW, Heeger AJ, Shirakawa H, Louis EJ, MacDiarmid AG (1978) J. Chem. Phys. 69: 5098
8. Nigrey PJ, MacDiarmid AG, Heeger AJ: J. Chem. Soc., Chem. Commun. 1979: 594
9. MacInnes D Jr, Druy MA, Nigrey PJ, Nairns DP, MacDiarmid AG, Heeger AJ: J. Chem. Soc., Chem. Commun. 1981: 317
10. Noufi R, Frank AJ, Nocic AJ (1981) J. Am. Chem. Soc. 103: 1849
11. Audebert P, Bidan G, Lapkowski M: J. Chem. Soc., Chem. Commun. 1986: 887
12. Thackeray JW, White HS, Wrighton MS (1985) J. Phys. Chem. 89: 5133
13. Albery WJ, Hillman AR (1981) Annu. Rep. Prog. Chem. Sect. C 78: 377
14. Murray RW (1984) in: Electroanalytical chemistry, Bard AJ, (ed) M. Dekker, New York, vol 13 p 191
15. Ryan MD, Wilson GS (1982) Anal. Chem. 54: 20R
15a. Merz A: Topics Curr. Chem., this volume
16. Proceedings of the International Conference on Low-Dimensional Conductors, Boulder, Colorado, August 1981: Mol. Cryst. Liq. Cryst. (1981) vol 77; (1982) vols 79, 81, 83, 85
17. La Physique et la Chimie des Polymeres Conductuers, Les Arcs, France, December 1983: J. Phys. Colloq. (1983) vol 44 C3
18. Symposium-Conducting organic polymers in energy conservation and storage, The Electrochemical Society Meeting, San Francisco, May 1983
19. Proceedings of the Workshop on Synthetic Metals (1984) Synth. Met. 9: 128
20. Proceedings of the International Conference on the Physics and Chemistry of Low-Dimensional Synthetic Metals, Abbano Terme, Italy, June 1984: Mol. Cryst., Liq. Cryst. (1985) vols 117–121
21. Proceedings of the International Conference on Science and Technology of Synthetic Metals, Kyoto, Japan, June 1986: Synth. Metals (1987) vols 17, 18
22. Proceedings of the International Conference on Electronic Processes in Conducting Polymers in Vadstena, Sweden, August 1986: Synth. Met. (1987) vol 21
23. Seymour RB (ed) (1981) Conductive polymers, Plenum, New York
24. Mort J, Pfister G (eds) Electronic properties of polymers, Wiley, New York
25a. Kuzmany H, Mehring M, Roth S (eds) (1985) Springer, Berlin
25b. Kuzmany H, Mehring M, Roth S, (eds) (1987) Electronic properties of conjugated polymers, Springer, Berlin (1986)
26. Skotheim (ed) (1986) Heidelberg New York, Handbook of conducting polymers, Marcel Dekker, New York

27. Frommer JE, Chance RR (1986) in: Electrically conductive polymers, in: Grayson M, Kroschwitz J (eds) Encyclopedia of Polymer Science and Engineering, 2nd edn, Wiley, New York, vol 5 p 462
28. Mair HJ, Roth S (eds) (1986) Elektrisch leitende Kunststoffe, Hanser, Munich
29. Wegner G (1981) Angew. Chem. 93: 352; (1981) Angew. Chem. Int. Ed. Engl. 20: 361
30. Greene RL, Street GB (1984) Science 226: 651
31a. Bruce M, Murphy L (1984) Nature 309: 119
31b. Hanack M (1983) Chimica 37: 238
32a. Chandler GK, Pletcher D (1985) Electrochemistry, Royal Soc. of Chemistry, London, vol 10 p 117
32b. Przyluski J, Roth S (eds) (1987) Electrochemistry of conducting polymers, Trans. Tech. Publications, Basel
33. Menke K, Roth S (1986) Chem. Unserer Zeit 20: 1, 33
34. Letheby H (1962) J. Chem. Soc. 15: 161
35. Goppelsroeder F (1976) Compt. rend. 82: 331, 1392
36. Dall'Olio A, Dascola Y, Varacca V, Bocchi V (1968) C.R. Hebd. Seances. Acad. Sci., Ser. C 267: 433
37. Lund H (1961) Elektrodenreaktioner i Organsik Polarografi og Voltammetri, Aarhus Stiftsbogtrykkorie, Aarhus
38a. Diaz AF, Kanazawa KK, Gardini GP: J. Chem. Soc., Chem. Commun. 1979: 635
38b. Kanazawa KK, Diaz AF, Geiss RH, Gill WD, Kwak JF, Logan JA, Rabolt JF, Street GB: J. Chem. Soc., Chem. Commun, 1979: 854
38c. Kanazawa KK, Diaz AF, Gill WD, Grant PM, Street GB, Gardini GP, Kwak JK (1979/80) Synth. Met. 1: 329
39. Angeli A (1916) Gazz. Chim. Ital. 46: 279
40. Gardini GP (1973) Adv. Heterocycl. Chem. 15: 67
41. Kovacic P, Jones MB (1987) Chem. Rev. 87: 357
42. Tourillon G, Garnier F (1982) J. Electroanal. Chem. 135: 173
43. Delamar M, Lacaze P-C, Dumousseau J-Y, Dubois, J-E (1982) Electrochim. Acta 27: 61
44. Brilmyer G, Jasinki R (1982) J. Electrochem. Soc. 129: 1950
45. Bryce MR (1985) Annu. Rep. Prog. Chem., Sect. B 82: 377
46. Chen S-A, Shy H-J (1985) J. Polym. Sci., Polym. Chem. 23: 2441
47. Satoh M, Uesugi F, Tabata M, Kaneto K, Yoshino K: J. Chem. Soc., Chem. Commun., 1986: 550
48. Satoh M, Uesugi F, Tabata M, Kaneto K, Yoshino K: J. Chem. Soc., Chem. Comm. 1986: 979
49. Bargon J, Mohamand S, Waltman RJ (1983) IBM. J. Res. Develop. 27: 330
50. Shacklette LW, Chance RR, Ivory DM, Miller GG, Baughman RH (1979) Synth. Met. 1: 307
51. Satoh M, Kaneto K, Yoshino K: J. Chem. Soc., Chem. Commun. 1985: 1629
52. Diaz AF, Logan JA (1980) J. Electroanal. Chem. 111: 111
53. Rault-Berthelot J, Simonet J (1986) Nouv. J. Chim. 10: 169
54. Waltman RJ, Bargon J (1985) J. Electroanal. Chem. 194: 49
55a. Wudl F, Kobayashi M, Heeger AJ (1984) J. Org. Chem. 49: 3381
55b. Wudl F, Kobayashi M, Colaneri N, Boysel M, Heeger AJ (1985) Mol. Cryst. Liq. Cryst. 118: 199
55c. Kobayashi M, Colanari N, Boysel M, Wudl F, Heeger AJ (1985) J. Chem. Phys. 82: 5717
56. Di Marco P, Mastragostino M, Taliani C (1985) Mol. Cryst. Liq. Cryst. 118: 241
57a. Lazzaroni R, Dujardin S, Riga J, Verbist J, Bredas JL, Delhalle J, Andre JM (1985) in: Kuzmany H, Mehring M, Roth S (eds) Electronic properties of polymers and related compounds, Springer, Berlin Heidelberg New York, p 191
57b. Lazzaroni R, de Prijck A, Riga J, Verbist J, Bonhomme C, Brose F, Christiaens L, Renson M (1987) Synth. Met. 18: 123
57c. Lazzaroni R, Riga J, Verbist J, Christiaens L, Renson M: J. Chem. Soc., Chem. Commun. 1985: 999
58. Danieli R, Ostja P, Tiecco M, Zamboni R, Taliani C: J. Chem. Soc., Chem. Commun. 1986: 1473
59. Sato M-A, Tanaka S, Kaeriyama K: J. Chem. Soc., Chem. Commun. 1986: 873

60. Gagnon DR, Capstran JD, Karasz FE, Lenz R, Antoun S (1987) Polymer 28: 567
61. Mazur S, Lugs, PS, Yarnitzky, C (1987) J. Electrochem. Soc. 134: 346
62. Labes MM, Loves P, Nichols LF (1979) Chem. Rev. 79: 1
63. Bhadani SN, Parravano G (1983) in: Baizer MM, Lund H (eds) Organic electrochemistry, M. Dekker, New York, p 995
64. Diaz AF, Castillo JI, Logan JA, Lee W-Y (1982) J. Electroanal. Chem. 129: 115
65. Diaz, AF (1981) Chem. Scr. 17: 145
66. Waltman RJ, Bargon J, Diaz AF (1983) J. Phys. Chem. 87: 1459
67. Street GB, Clarke TC, Krounbi M, Kanazawa KK, Lee V, Pfluger P, Scott JC, Weiser G (1982) Mol. Cryst. Liq. Cryst. 83: 253
68. Yamamoto T, Sanechika K, Yamonoto A (1980) J. Poly. Sci., Polym. Lett. Ed. 18: 9
69. Diaz AF, Martinez A, Kanazawa KK, Salmon M (1980) J. Electroanal. Chem. 130: 181
70. Street GB (1986) in: Skotheim TA (ed) Handbook of conducting polymers, Dekker, New York, p 265
71. Bard AJ, Ledwith A, Shine HJ (1976) Adv. Phys. Org. Chem. 12: 155
72. Hammerich O, Parker VD (1984) Adv. Phys. Org. Chem. 20: 55
73. Genies EM, Bidan G, Diaz AF (1983) J. Electroanal. Chem. 149: 101
74. Inoue T, Yamase T (1983) Bull. Chem. Soc. Jpn. 56: 985
75a. Kossmehl G, Chatzitheodorou G. (1982) Makromol. Chem. Rapid. Commun. 2: 551
75b. Kossmehl G, Chatzitheodorou G (1982) Mol. Cryst. Liq. Cryst. 83: 291
76. Asavapiriyanont S, Chandler GK, Gunawardena GA, Pletcher D (1984) J. Electroanal. Chem. 177: 229
77a. Heinze J, Hinkelmann K, Dietrich M, Mortensen J (1986) DECHEMA Monographie 102: 209
77b. Heinze J, Hinkelmann K, Land M (1989) „Organische Elektrochemie — Angewandte Elektrothermie", DECHMA Monographie Vol 112, VCH, Weinheim, p. 75
78. Enkelmann V, Morra BS, Kröhnke C, Wegner G, Heinze J (1982) Chem. Phys. 66: 303
79. Aalstad B, Ronlan A, Parker VD (1981) Acta Chem. Scand. B35: 649
80. Parker VD (1983) Adv. Phys. Org. Chem. 19: 131
81. Amatore C, Saveant JM (1983) J. Electroanal. Chem. 144: 59
82. Eigen M, Kruse W, Maas G, De Maeyer L (1964) in: Porter C (ed) Progress in reaction kinetics, Pergamon, Oxford, vol 2 p 284
83. Debye P (1942) Trans. Electrochem. Soc. 82: 265
84. Heinze J, Hinkelmann K, Dietrich M, Mortensen J (1985) Ber. Bunsenges. Phys. Chem. 89: 1225
85. Ambrose JF, Nelson RF (1968) J. Electrochem. Soc. 115: 1159
86. Ambrose JF, Carpenter LL, Nelson RF (1975) J. Electrochem. Soc. 122: 876
87. Lamm W, Pragst F, Jugelt W (1975) J. prakt. Chem. 317: 995
88. Street GB, Lindsey SE, Nazzal AI, Wynne KJ (1985) Mol. Cryst. Liq. Cryst. 118: 137
89. Downard AJ, Pletcher D (1986) J. Electroanal, Chem. 206: 139
90. Diaz AF, Hall B (1983) IBM J. Res. Develop. 37: 342
91. Prejza J, Lundström I, Skotheim TA (1982) J. Electrochem. Soc. 128: 1685
92. Downard AJ, Pletcher D (1986) J. Electroanal. Chem. 206: 146
93. Osa T, Yildiz A, Kuwana T (1969) J. Am. Chem. Soc. 81: 3994
94. Dietrich M, Mortensen J, Heinze J: J. Chem. Soc., Chem. Commun. 1986: 1131
95. Wernet W, Monkenbusch M, Wegner G (1984) Makromol. Chem., Rapid Commun. 5: 1574
96. Wernet W, Monkenbusch M, Wegner G (1985) Mol. Cryst. Liq. Cryst. 118: 193
97. Tabakubo M (1987) Synth. Met. 18: 53
98. Warren LF, Andersen DP (1987) J. Electrochem. Soc. 134: 101
99. Genies EM, Syed AA (1984) Synth. Met. 10: 21
100. Salmon M, Carbajal ME, Juarez JC, Diaz AF, Rock MC (1984) J. Electrochem. Soc. 131: 1802
101. Diaz AF, Castillo J, Kanazawa KK, Logan JA, Salmon M, Fajardo O (1982) J. Electroanal. Chem. 133: 233
102. Audebert P, Bidan G (1985) Mol. Cryst. Liq. Cryst. 118: 187
103. Salmon M, Carbajal ME, Aguilar M, Saloma M, Juarez JC: J. Chem. Soc., Chem. Commun. 1983: 1532
104a. Ritchie CD, Sager WF (1964) Prog. Phys. Org. Chem. 2: 323
104b. Zuman P (1967) Substituent effects in organic polarography, Plenum, New York

105. Waltman RJ, Diaz AF, Bargon J (1984) J. Phys. Chem. 88: 4343
106a. Waltman RJ, Bargon J (1984) Can. J. Chem. 64: 76
106b. Waltman RJ, Bargon J (1984) Tetrahedron 40: 3963
107. Asavapiriyanont S, Chandler GK, Gunawardena GA, Pletcher D (1984) J. Electroanal. Chem. 177: 245
108. Downard AJ, Pletcher D (1986) J. Electroanal. Chem. 206: 139
109. Pickup PG, Osteryoung RA (1984) J. Am. Chem. Soc. 106: 2294
110. Hillman AR, Mallen EF (1987) J. Electroanal. Chem. 220: 351
111. Thyssen A, Borgerding A, Schultze JW (1987) Makromol. Chem. Macromol. Symp. 8: 143
112. Southampton Electrochemistry Group (1985) Instrumental methods in electrochemistry, Ellis Horwood, Chichester
113. Lang P, Chao F, Costa M, Garnier F (1987) Polymer 28: 668
114. Fleischmann M, Thirsk HR (1963) in: Delahay P (ed) Advances in electrochemistry and electrochemical engineering, Wiley-Interscience, New York, vol 3 p 123
115. Schultze JW, Lohrengel MM, Roß D (1978) Electrochim. Acta 28: 973
116. Hillmann AR (1987) in: Linford R (ed) Electrochemical technology of polymers, Elsevier, London, p 103
117. Clarke TC, Scott JC, Street GB (1983) IBM J. Res. Dev. 27: 313
118. Street GB (1986) in: Skotheim TA (ed) Handbook of conducting polymers, Dekker, New York, p 265
119a. Pfluger P, Krounbi M, Street GB, Weiser G (1983) J. Chem. Phys. 78: 3212
119b. Pfluger P, Street GB (1984) J. Chem. Phys. 80: 544
120. Geiss RH, Street GB, Volksen W, Economy J (1983) IBM J. Res. Dev. 27: 321
121. Hotz CZ, Kovacic P, Khoury IA (1983) J. Polym. Sci. Polym. Chem. Ed. 21: 2617
122. Yamamoto T, Sanechika K, Yamomoto A (1983) Bull. Chem. Soc. Jpn. 56: 1497
123. Neugebauer H, Neckel A, Brinda-Konopik N (1985) in: Kuzmany H et al. (eds) Electronic properties of polymers and related compounds, Springer, Berlin Heidelberg New York, p 227
124. Tourillon G, Garnier F (1983) J. Phys. Chem. 87: 2289
125. Neugebauer H, Nauer G, Neckel A, Tourillon G, Garnier F, Lang P (1984) J. Phys. Chem. 88: 652
126. Hotta H, Hosaka T, Shimotsuma W (1984) J. Chem. Phys. 80: 954
127. Ford WK, Dike CB, Salaneck WR (1982) J. Chem. Phys. 77: 5030
128. Ritsko JJ, Fink J, Crecelius G (1983) Solid State Commun. 46: 477
129. Nazzal A, Street GB: J. Chem. Soc. Chem. Commun. 1983: 83
130. Sato M, Kaeriyama K, Someno K (1983) Makromol. Chem. 184: 2241
131. Hotta S, Rughooputh SDDV, Heeger AJ, Wudl F (1987) Macromolecules 20: 212
132. Blankespoor R, Miller LL: J. Chem. Soc., Chem. Commun. 1985: 90
133. Chang AC, Blankespoor RL, Miller LL (1987) J. Electroanal. Chem. 236: 239
134. Hubbard AT, Anson FC (1970) J. Electroanal. Chem. 4: 129
135. Lane RF, Hubbard AT (1978) J. Phys. Chem. 77: 1401
136. Laviron E (1979) J. Electroanal. Chem. 100: 263
137. Angerstein-Kozlowska H, Klinger J, Conway BE (1977) J. Electroanal. Chem. 75: 45, 61
138. Laviron E (1974) J. Electroanal. Chem. 52: 355, 395
139. Hubbard AT (1968) J. Electroanal. Chem. 22: 165
140. Angerstein-Kozlowska H, Conway BE, Klinger J (1978) J. Electroanal. Chem. 87: 301, 321
141. Kaufmann FB, Schroeder AH, Engler EM, Kramer SR, Chambers JQ (1980) J. Am. Chem. Soc. 102: 483
142. Daum P, Lenhard JR, Rolison DR, Murray RW (1980) J. Am. Chem. Soc. 102: 4649
143. Daum P, Murray RW (1981) J. Phys. Chem. 85: 389
144. Doblhofer K, Braun K, Lange R (1986) J. Electroanal. Chem. 206: 93
145. Garnier F, Tourillon G, Gazard M, Dubois JC (1983) J. Electroanal. Chem. 148: 299
146. Tourillon G (1986) in: Skotheim TA (ed) Handbook of Conducting Polymers, M. Dekker, New York
147. Brédas JL, Silbey R, Boudreaux DS, Chance RR (1983) J. Am. Chem. Soc. 105: 6555
148. Diaz AF, Crowley J, Bargon J, Gardini GP, Torrance JB (1981) J. Electroanal. Chem. 121: 355
149. Nigrey, PJ, MacDiarmid AG, Heeger, AJ (1982) Mol. Cryst. Liq. Cryst. 83: 309

150. Tourillon G, Garnier F (1984) J. Electroanal. Chem. 161: 51
151. Deuchert K, Hünig S (1978) Angew. Chem. 90: 927; (1978) Angew. Chem. Int. Ed. Engl. 17: 875
152a. Hörhold H-H, Helbig M, Raabe D, Räthe H, Opfermann J (1984) Wiss. Ber. Akad. Wiss. DDR., Zentralinstitut. Festkörperphys. Werkstofforsch. 29: 87
152b. Hörhold H-H, Opfermann J, Atrat P, Tauer K-D, Drefahl G: 5th Internat. Symp. Polykondensation, Varna 1975, in "Polykondensationsprozesse", Sofia 1976, Verlag der Bulg. Akad. Wiss., p. 171
152c. Hörhold H-H, Helbig M, Raabe D, Opfermann J, Scherf U, Stockmann R, Weiß D (1987) Z. Chem. 27: 126
153. Genies EM, Pernaut J-M (1985) J. Electroanal. Chem. 191: 111
154. Brazovskii SA, Kirova N (1981) JEPT Lett. 33: 4
155. Brédas JL, Chance RR, Silbey R (1981) Mol. Cryst. Liq. Cryst. 77: 319; (1982) Phys. Rev. B26: 5843
156. Brédas JL, Thémans B, André JM (1983) Phys. Rev. B27, 7827; (1983) J. Chem. Phys. 78: 6137
157. Fesser K, Bishop AR, Campbell DK (1983) Phys. Rev. B27: 4804
158. Brédas JL (1985) Mol. Cryst. Liq. Cryst. 118: 49
159. Brédas JL, Street GB (1985) Acc. Chem. Res. 18: 308
160. Brédas JL, Scott JC, Yakushi K, Street GB (1984) Phys. Rev. B30: 1023
161. Scott JC, Pfluger P, Krounbi MT, Street GB (1983) Phys. Rev. B28: 2140
162. Scott JC, Brédas JL, Yakushi K, Pfluger P, Street GB (1984) Synth. Met. 9: 165
163a. Scott JC, Brédas JL, Kaufmann JH, Pfluger P, Street GB, Yakushi K (1985) Mol. Cryst. Liq. Cryst. 118: 163
163b. Kaufmann JH, Colaneri N, Scott JC, Kanazawa KK, Street GB (1985) Mol. Cryst. Liq. Cryst. 118: 171
164a. Heinze J, Serafimov O, Zimmermann HW (1974) Ber. Bunsenges. Phys. Chem. 78: 652
164b. Heinze J, Zimmermann HW (1977) Ber. Bunsenges. Phys. Chem. 78: 321
165. Dister D, Hohlneicher G (1970) Ber. Bunsenges. Phys. Chem. 74: 960
166. Hoijtink GJ, Zandstra PJ (1960) Mol. Phys. 3: 371
167. Brédas JL, Thémans B, André JM, Chance RR, Silbey R (1985) Synth. Met. 9: 265
168. Brédas JL, Thémans B, Fripiat JG, André JM, Chance RR (1984) Phys. Rev. B29: 6761
169. Nechtschein M, Devreux F, Genould F, Vieil E, Pernaut, JM, Genies E (1986) Synth. Met. 15: 59
170. Genoud F, Guglielmi M, Nechtschein M, Genies E, Salmon M (1985) Phys. Rev. Lett. 55: 118
171. Bull RA, Fan FF, Bard AJ (1982) J. Electrochem. Soc. 129: 1009
172. Jakobs RCM, Janssen LJJ, Barendrecht E (1984) Rec. Trav. Chim. Pays-Bas 103: 275
173. Feldberg SW (1984) J. Am. Chem. Soc. 106: 4671
174. Tanguy J, Mermilliod N, Hoclet M (1987) J. Electrochem. Soc. 134: 795
175. Heinze J, Dietrich M, Mortensen J (1987) Makromol. Chem., Macromol. Symp. 8: 73
176. Trinidad F, Alonso-Lopez J, Nebot M (1987) J. Appl. Electrochem. 17: 215
177. Beck F, Oberst M (1987) Makromol. Chem., Macromol. Symp. 8: 87
178. Bittihn R, Ely G, Woeffler F, Münstedt H, Naarmann H, Naegele D (1987) Makromol. Chem., Macromol. Symp. 8: 51
179. Salmon M, Diaz AF, Logan AJ, Krounbi M, Bargon J (1982) Mol. Cryst. Liq. Cryst. 83: 1297
180. Hotta S, Hosaka T, Shimotsuma W, Synth. Met. (1983) 6: 69
181. Chung TC, Kaufman JH, Heeger AJ, Wudl F (1984) Phys Rev. 30B: 702
182. Kaneto K, Yoshino K, Inuishi Y (1983) Jpn. J. Appl. Phys. 22: L 412, L 567
183a. Shacklette LW, Elsenbaumer RL, Baughman RH (1983) J. Phys. Coll. 44 C3: 559
183b. Shacklette LW, Elsenbaumer RL, Chance RR, Sowa JM, Ivory DM, Miller GG, Baughman RH: J. Chem. Soc. Chem. Commun. 1982: 361
184. Elsenbaumer RL, Shacklette LW (1986) in: Skotheim TA (ed) Handbook of conducting polymers, M. Dekker, New York, p 213
185. Maurice F, Froyer G, Pelous Y (1983) J. Phys. Colloq. 44 C3: 587
186. Mac Diarmid AG, Yang LS, Huang WS, Humphrey BD (1987) Synth. Met. 18: 393
187. Beck F, Oberst M, Braun B (1987) DECHEMA Mon. 109: 457

188. Baughman RH, Delanoy P, Murthy NS, Miller GG, Eckhardt H, Shacklette LW (1984) Mol. Cryst. Liq. Cryst. 106: 415
189. Maxfield M, Mu SL, MacDiarmid AG (1985) J. Electrochem. Soc. 132: 838
190. MacDiarmid AG, Kaner RB (1986) in: Skotheim TA (ed) Handbook of conducting polymers, M. Dekker, New York p 689
191. MacDiarmid AG, Maxfield MR (1987) in: Linford RG (ed) Electrochemical science and technology of polymers-1, Elsevier Applied Science, London, p 67
192. Genies EM, Pernault, JM (1984/85) Synth. Met. 10: 117
193. Kaufman JH, Kanazawa KK, Street GB (1984) Phys. Rev. Lett. 53: 2461
194. Chao F, Baudoin JL, Costa M, Lang P (1987) Makromol. Chem., Macromol. Symp. 8: 173
195. Zhou Q-X, Kolaskie CJ, Miller LL (1987) J. Electroanal. Chem. 223: 283
196. Heinze J, Mortensen J, Müllen K, Schenk R: J. Chem. Soc., Chem. Commun. 1987: 701
197a. Heinze J, Störzbach M, Mortensen J (1987) Ber. Bunsenges. Phys. Chem. 91: 960
197b. Heinze J, Mortensen J, Störzbach M (1987) in: Kuzmany H, Mehring M, Roth S (eds) Electrochemical properties of conjugated polymers, Springer, Berlin Heidelberg New York, p 385
198. Nigrey RJ, Heeger AJ, MacDiarmid AG (1982) Mol. Cryst. Liq. Cryst. 83: 309
199a. Müller F, Meisterle P, Kuzmany H (1985) Mol. Cryst. Liq. Cryst. 121: 237
199b. Kuzmany H (1985) Pure Appl. Chem. 57: 235
200. El-Khodary A, Bernier P (1985) Mol. Cryst. Liq. Cryst. 117: 127
201. Feldblum A, Kaufman JH, Etemad S, Heeger AJ, Chung TC, MacDiarmid AG (1982) Phys. Rev. B 26: 815
202. Evans DH, O'Connell KM (1986) in: Bard AJ (ed) Electroanalytical Chemistry, M. Dekker, New York, vol 14 p 113
203. Dietrich M, Heinze J, Fischer H, Neugebauer FA (1986) Angew. Chem. 98: 999; (1986) Angew. Chem. Int. Ed. Engl. 25: 1021
204. Thompson AJ (1978) Phys. Rev. Lett. 40: 1511
205. Kaufman JH, Kaufer JW, Heeger AJ, Kaner RB, MacDiarmid AG (1982) Phys. Rev. B26: 2327
206. Kaufman JH, Chung TC, Heeger AJ (1984) J. Electrochem. Soc. 134: 2847
207. Heeger AJ (1985) Phil. Trans. R. Soc. Lond. A 314: 17
208. Chung TC, Kaufman JH, Heeger AJ (1984) Phys. Rev. B 30: 702
209. Scott JC, Brédas JL, Yakushi K, Street GB (1984) Phys. Rev. B 30: 1023
210. Kitani A, Izumi J, Yano J, Hiromoto Y, Susaki K (1984) Bull. Chem. Soc. Jpn. 57: 2254
211. Huang W-S, Humphrey BD, MacDiarmid AG (1986) J. Chem. Soc., Faraday. Trans. 1, 82: 2385
212. Heinze J, Mortensen J, Hinkelmann K (1987) Synth. Met. 21: 209
213. Breitenbach M, Heckner KH (1973) J. Electroanal. Chem. 43: 267
214. Noufi R, Nozik AJ, White J, Warren LF (1981) J. Electrochem. Soc. 129: 2261
215. Kobayashi T, Yaneyama H, Tamura H (1984) J. Electroanal. Chem. 161: 419
216. MacDiarmid AG, Chiang C, Halpern M, Huang W-S, Mu S-L, Somasiri NLD, Wu W, Yaniger SI (1985) Mol. Cryst. Liq. Cryst. 121: 173
217. Genies EM, Tsintavis C (1986) J. Electroanal. Chem. 200: 127
218. DeSurville R, Jozefowicz M, Yu LT, Perichon J, Buvet R (1968) Electrochim. Acta 13: 1451
219. Dunsch L (1975) J. Electroanal. Chem. 61: 61; J. prakt. Chem. 317: 409
220. Chiang JC, MacDiarmid AG (1986) Synth. Met. 13: 193
221. Wudl F, Angus RO Jr, Lu FL, Allemand PM, Vachon DJ, Nowak M, Liu ZX, Heeger AJ (1987) J. Am. Chem. Soc. 109: 3677
222. Boudreaux DS, Chance RR, Wolf JF, Shacklette LW, Brédas JL, Thémans B, André JM, Silbey R (1986) J. Chem. Phys. 85: 4584
223. Genies EM, Lapkowski M (1987) J. Electroanal. Chem. 220: 67; (1987) in: Kuzmany H, Mehring M, Roth S (eds) Electronic properties of conjugated polymers, Springer, Berlin Heidelberg New York, p 223; (1987) Synth. Met. 21: 117
224a. Sariciftci NS, Neugebauer H, Kuzmany H, Neckel A (1987) in: Kuzmany H, Mehring M, Roth S (eds) Electronic properties of conjugated polymers, Springer, Berlin Heidelberg New York, p 228
224b. Sariciftci NS, Kuzmany H, Neugebauer H (1987) J. Mol. Electron.

225a. Duke CB (1987) Synth. Met. 21: 5
225b. Roth S (1987) Synth. Met. 21: 51
225c. MacDiarmid AG (1987) Synth. Met 21: 79
226. Wirsén A (1987) Electroactive polymer materials, Technomic, Basel
227. Kaneko M, Wöhrle D (1988) in: Cantow H-J (ed) Advances in polymer sciences 84, Springer, Berlin Heidelberg New York, p 141
228. Potember RS, Hoffman RC, Hu HS, Cocchiaro JE, Viands CA, Murphy RA, Poehler TO (1987) Polymer 28: 574
229. Gazard M (1986) in: Skotheim TA (ed) Handbook of conducting polymers, M. Dekker, New York, p 673
230. Meyer WH, Kiess H, Binggelli B, Meier E, Harbeke G (1985) Synth. Met. 10: 255
231. Duke CB, Gibson HW (1982) in: Kirk-Othmer J (ed) Encyclopedia of chemical Technology, Wiley, New York, p 755
232. DePaoli M-A, Waltman RJ, Diaz AF, Bargon J: J. Chem. Soc., Chem. Commun. 1984: 1015
233. Noufi R, Tench D, Warren LF (1981) J. Electrochem. Soc. 128: 2596
234. Soga K, Ikeda S (1986) in: Skotheim TA (ed) Handbook of conducting polymers, M. Dekker, New York, p 661
235. Nigrey PJ, MacInnes D, Nairns P MacDiarmid AG, Heeger AJ (1981) J. Electrochem. Soc. 128: 1651
236. Kaneto K, Maxfield M, Nairns P, MacDiarmid AG (1982) J. Chem. Soc., Faraday Trans. 1, 78: 3417
237. Kaner RB, MacDiarmid AG (1984) J. Chem. Soc., Faraday Trans. 1, 80: 2109
238. Farrington GC, Scrosati B, Frydrych D, De Nazzio J (1984) J. Electrochem. Soc. 131: 7
239. Mu SL, Porter SJ, Wu W, MacDiarmid AG (1984) J. Electrochem. Soc. 131: 7
240. Nagotomo T, Kakehata H, Ichikawa C, Omoto O (1985) J. Electrochem. Soc. 132: 1380
241. Caja J, Kaner RB, MacDiarmid AG (1984) J. Electrochem. Soc. 131: 2744
242. Will FG (1985) J. Electrochem. Soc. 132: 2351
243. Beniere F, Boils D, Canepa H, Franco J, Le Corre A, Louboutin JP (1985) J. Electrochem. Soc. 132: 2100
244. Huq R, Farrington GC (1985) J. Electrochem. Soc. 132: 1432
245. Shacklette LW, Toth JE, Murthy NS, Baughman RH (1985) J. Electrochem. Soc. 132: 1529
246. Maxfield M, Wolf JF, Miller GG, Frommer JE, Shacklette LW (1986) J. Electrochem. Soc. 133: 117
247. Broich B, Hocker J (1984) Ber. Bunsenges. Phys. Chem. 89: 497
248. Dietz KH, Beck F (1985) J. Appl. Electrochem. 15: 159
249. Nagatomo T, Ichikawa C, Omoto O (1987) J. Electrochem. Soc. 134: 305
250. Kaner RB, Porter GJ, MacDiarmid AG (1986) J. Chem. Soc., Faraday Trans. I, 82: 2323
251. Kaner RB, MacDiarmid AG (1986) Synth. Met. 14: 3
252. Chiang CK (1981) Polym. Commun. 22: 1454
253. Nagotomo T, Kakehata H, Ichikawa C, Omoto O (1985) Jpn. J. Appl. Phys. 24: L 397
254. Naarman H, Theophilou N (1987) Synth. Met. 22: 1
255. Wieners G, Monkenbusch M, Wegner G (1984) Ber. Bunsenges. Phys. Chem. 88: 935
256. Pickup PG, Osteryoung RA (1985) J. Electroanal. Chem. 195: 271
257. Mermilliod M, Tanguy J, Petiot F (1986) J. Electrochem. Soc. 133: 1073
258. Mohammadi A, Ingañas O, Lundström I (1986) J. Electrochem. Soc. 133: 947
259a. Osaka T, Naoi K, Sakai H, Ogano S (1987) J. Electrochem. Soc. 134: 285
259b. Osaka T, Naoi K, Ogano S, Nakamura S (1987) J. Electrochem. Soc. 134: 2096
260. Kaufman JH, Chung TC, Heeger AJ, Wudl F (1984) J. Electrochem. Soc. 131: 2092
261. Nogami T, Nawa M, Mikawa HJ: J. Chem. Soc., Chem. Commun. 1982: 1158
262. Biserni M, Marinangeli A, Mastragostino M (1986) J. Electrochem. Soc. 132: 1597
263a. Hirabayashi T, Naoi K, Osaka T (1987) J. Electrochem. Soc. 134: 758
263b. Kakuta T, Shirota Y, Mikawa H: J. Chem. Soc., Chem. Comun. 1985: 553
264a. Papir YS, Karkov VP, Current SP (1983) Electrochem. Soc., Ext. Abstr. 83: 820
264b. Schroeder AH, Papir YS, Karkov VP (1983) Electrochem. Soc., Ext. Abstr. 83: 822
265. Kitani A, Kaya M, Sasaki K (1986) J. Electrochem. Soc. 133: 1069
266. Genies EM, Lapkowski M, Santier C, Vieil E (1987) Synth. Met. 18: 631
267. MacDiarmid AG, Yang LS, Huang W.-W, Humphrey BD (1987) Synth. Met. 18: 393

268. Travers JB, Chroboczek J, Devreux F, Genoud F, Nechtschein M, Syed A, Genies EM, Tsintavis C (1985) Mol. Cryst. Liq. Cryst. 121: 195
269. Kobayashi T, Yoneyama H, Tamura H (1984) J. Electroanal. Chem. 177: 293
270. Beck F, Pruss A (1987) J. Electroanal. Chem. 216: 157
271. Münstedt H (1986) Polymer 27: 899
272. Beck F, Braun P, Oberst F (1987) Ber. Bunsenges Phys. Chem. 91: 967
273. Ingañs O, Lundström I (1987) Synth. Met. 21: 13
274. Kuwahata S, Yoneyama H, Tamura H (1984) Bull. Chem. Soc. Jpn. 57: 2247
275. Yoneyama H, Wakomoto K, Tamura H (1985) J Electrochem. Soc. 132: 2414
276. Yosino K, Kaneto K, Inuishi Y (1983) Jpn. J. Appl. Phys. 22: L 157
277. Kobayashi T, Yoneyama H, Tamura H (1984) J. Electroanal. Chem. 161: 419; (1984) 177: 281
278. Kitani A, Yano J, Sasaki K (1986) J. Electroanal. Chem. 209: 227
279. Kaneko M, Nakamura H (1987) Makromol. Chem., Rapid Commun. 8: 179
280. Yashima H, Kobayashi M, Lee K-B, Chung D, Heeger AJ, Wudl F (1987) J. Electrochem. Soc. 134: 46
281. Wang TT, Tasaka S, Hutton RS, Lu PY: J. Chem. Soc., Chem. Comun. 1985: 1343
282. DePaoli M-A, Waltman RJ, Diaz AF, Bargon J (1985) J. Poly. Sci., Poly. Chem. Ed. 23: 1687
283. Niwa O, Tamamura T: J. Chem. Soc., Chem. Commun. 1984: 817
284. Niwa O, Hikita M, Tamamura T (1985) Makromol. Chem., Rapid. Commun, 6: 375
285. Niwa O, Hikita M, Tamamura T (1985) Polym. Prepr. Jpn. 34: 2821
286. Lindsey SE, Street GB (1984/85) Synth. Met. 10: 67
287. Lindenberger H, Roth S, Hanack M (1985) in: Kuzmany H, Mehring M, Roths (eds) Electronic properties of polymers and related compounds, Springer, Berlin Heidelberg New York p 194
288. Penner RM, Martin CR (1986) J. Electrochem. Soc. 133: 310
289. Fan F-RF, Bard AJ (1986) J. Electrochem. Soc. 133: 301
290. Nagasubramanian G, DiStefano S, Moacanin J (1986) J. Phys. Chem. 90: 4447
291. Nazzal AI, Street GB: J. Chem. Soc., Chem. Commun. 1985: 375
292. Druy MA (1986) J. Electrochem. Soc. 133: 353
293. Roncali J, Garnier F: J. Chem. Soc., Chem. Commun. 1986: 783
294. Wessling B (1987) in: Kuzmany H, Mehring M, Roths (eds) Electronic properties of conjugated polymers, Springer, Berlin Heidelberg New York, p 407
295. Wnek GE (1986) in: Skotheim TA (ed) Handbook of conducting polymers, M. Dekker, New York, p 205
296. Ikeda O, Okabayashi K, Tamura H: Chem. Lett. 1983: 1821
297. Noufi R (1983) J. Electrochem. Soc. 130: 2126
298. Bull RA, Fan F-RF, Bard AJ (1983) J. Electrochem. Soc. 130: 1636
299. Skotheim TA, Rosenthal MV, Linkous CA: J. Chem. Soc., Chem. Commun. 1985: 612
300. Okabayashi K, Ikeda O, Tamura H: J. Chem. Soc., Chem. Commun. 1985: 684
301. Bedioui F, Bongars C, Devynck J, Bied-Charreton C, Hinnen C (1986) J. Electroanal. Chem. 207: 87
302. Cosnier S, Deronzier A, Moutet J-C (1985) J. Electroanal. Chem. 193: 193
303. Cosnier S, Deronzier A, Moutet J-C (1986) J. Electroanal. Chem. 207: 315
304. Daire F, Bedioui F, Devynck J, Bied-Charreton C (1986) J. Electroanal. Chem. 205: 309; (1987) 224: 95
305. Coche L, Dèronzier A, Moutet J-C (1986) J. Electroanal. Chem. 198: 187
306. Coche L, Moutet J-C (1987) J. Electroanal. Chem. 224: 111
307. Deronzier A, Latour J-M (1987) J. Electroanal. Chem. 224: 295
308. Audebert P, Bidan G, Lapkowski M, Limosin D (1987) in: Kuzmany H, Mehring M, Roth S (eds) Electronic properties of conjugated polymers, Springer, Berlin Heidelberg New York, p 366
309. Bidan G, Deronzier A, Moutet JC: J. Chem. Soc., Chem. Commun. 1984: 1185
310a. Haimerl A, Merz A (1986) Angew. Chem. 98: 179; (1986) Angew. Chem. Int. Ed. Engl. 25: 180
310b. Merz A, Baumann R, Haimerl A (1987) Makromol. Chem., Macromol. Symp. 8: 61
311. Shacklette LW, Elsenbaumer RL, Chance RR, Eckhardt H, Frommer JE, Baughman RH (1981) J. Chem. Phys. 75: 1919

312. Ikeda J, Ozaki M, Arakawa T: J. Chem. Soc., Chem. Commun. 1983: 1518
313. Brédas JL, Thémans B, André JM, Heeger AJ, Wudl F (1985) Synth. Met. 11: 344
314a. Elsenbaumer RL, Jen K-Y, Miller GG, Eckhardt H, Shacklette LW, Jow R (1987) in: Kuzmany H, Mehring M, Roth S (eds) Electronic properties of conjugated polymers, Springer, Berlin Heidelberg New York, p 400
314b. Jen K-Y, Eckhardt H, Jow TR, Shacklette LW, Elsenbaumer RL: J. Chem. Soc., Chem. Commun. 1988: 215
315. White HS, Kittlesen GP, Wrighton MS (1984) J. Am. Chem. Soc. 106: 5375
316. Kittlesen GP, White HS, Wrighton MS (1984) J. Am. Chem. Soc. 106: 7389
317. Paul EW, Ricco AJ, Wrighton MS (1985) J. Phys. Chem. 89: 1441
318. Koezuka H, Hyodo K, MacDarmid AG (1985) J. Appl. Phys. 58: 1279
319. Tsumura A, Koezuka H, Tsunoda S, Ando T: Chem. Lett. 1986: 683
320. Okano M, Fujishima A, Honda K (1985) J. Electroanal. Chem. 185: 393
321. Thackeray JW, Wrighton MS (1986) J. Phys. Chem. 90: 6674
322. Josowicz M, Janata J (1986) Anal. Chem. 58: 514
323. Cooper G, Noufi R, Frank AJ, Nozik AJ (1982) Nature 295: 578
324. Noufi R, Tench D, Warren LF (1980) J. Electrochem. Soc. 127: 2310
325. Skotheim TA, Lundström I, Prejza J (1981) J. Electrochem. Soc. 128: 1625
326. Skotheim TA, Lundström I, Delahoy AE, Kampas FJ, Vanier PE (1982) Appl. Phys. Lett. 40: 281
327. Fan F-RF, Wheeler BL, Bard AJ, Noufi R (1981) J. Electrochem. Soc. 128: 2042
328. Simon RA, Ricco AJ, Wrighton MS (1982) J. Am. Chem. Soc. 104: 2031
329. Skotheim TA, Ingañas O, Prejza J, Lundström I (1982) Mol. Cryst. Liq. Cryst. 83: 329
330. Frank AJ, Honda K (1982) J. Phys. Chem. 86: 1933; (1983) J. Electroanal. Chem. 150: 673
331. Skotheim TA, Ingañas O (1985) Mol. Cryst. Liq. Cryst. 121: 285
332. Rajeshwar K, Kaneki M, Yamada A (1983) J. Electrochem. Soc. 130: 38
333. Horowitz G, Tourillon G, Garnier F (1984) J. Electrochem. Soc. 131: 151
334. Horowitz G, Garnier F (1985) J. Electrochem. Soc. 132: 684
335. Garnier F, Horowitz G (1987) in: Kuzmany H, Mehring M, Roth S (eds) Electronic properties of conjugated polymers, Springer, Berlin Heidelberg New York, p 423
336. Schnöller M, Wersing W, Naarmann H (1987) Makromol. Chem., Macromol. Symp. 8: 83
337. Ingañas O, Lundström I (1984) J. Electrochem. Soc. 131: 1129
338. Kaneko M, Okuzumi K, Yamada A (1985) J. Electroanal. Chem. 183: 407
339. Inoue T, Yamase T (1981) Bull. Chem. Soc. Jpn. 54: 2817
340. Kaneko M, Nakamura HJ: J. Chem. Soc., Chem. Commun. 1985: 346
341. Kemochi T, Tsuchida E, Kaneko M, Yamada A (1985) Electrochim. Acta 30: 1405
342. Scrosati B (1988) Progress in Solid State Chemistry 18: 1
343. Ward MD (1989) in: Bard AJ (ed) Electroanalytical Chemistry, M. Dekker, New York, vol 16 p 181
344. Patil AO, Heeger AJ, Wudl F (1988) Chem. Rev. 88: 183
345. Proceedings of International Conference on Science and Technology of Synthetic Metals, Santa Fe, USA, June 1988: Synth. Met. (1989), Vols. 27–29
346. Tanaka S, Kaeriyama K, Hiraide T (1988) Makromol. Chem., Rapid. Commun. 9: 743
347. Zotti G, Schiavon G (1989) Makromol. Chem. 190: 405
348. Oyama N, Ohsaka T, Miyamoto H, Tanaka S, Kiyomine A, Kumagai T, Miyashi T, Mukai T (1989) Synth. Met. 28: C193
349. Bolognesi A, Catellani M, Destri S, Porzio W, Danieli R, Rossini S, Taliani C, Zamboni R, Ostoja P (1989) Synth. Met. 28: C521
350. Wudl F, Ikenoue Y, Patil AO in Ulrich D, Prassad PN (eds) Non-linear Optical and Electroactive Polymers, Plenum Press, New York, in press
351a. Bredas JL (1985) in: Kuzmany H, Mehring M, Roth S (eds) Electronic properties of polymers and related compounds, Springer, Berlin, Heidelberg, New York, p. 166
351b. Kertesz M, Lee YS (1989) Synth. Met. 28: C545
352. Yashima H, Kobayashi M, Lee KB, Chung D, Heeger AJ, Wudl F (1987) J. Electrochem. Soc. 134: 46
353. Elsenbaumer RL, Jen KY, Oboodi R (1986) Synth. Met. 15: 169
354. Sato M, Tanaka S, Kaeriyama K (1989) Synth. Met. 28: 229

355. Rughooputh SDDV, Nowak N, Hotta S, Heeger AJ, Wudl F (1987) Synth. Met. 21: 41
356. Roncali J, Garreau R, Yasser Y, Marque P, Garnier F, Lemaire M (1987) J. Phys. Chem. 91: 6706
357. Feldhues M, Kämpf G, Litterer H, Mecklenburg T, Wegener P (1989) Synth. Met. 28: C487
358. Imanishi K, Satoh M, Yasuda Y, Tsushima R, Aoki S (1988) J. Electroanal. Chem. 242: 203
359. Oberst M, Beck F (1987) Angew. Chem. Int. Ed. Engl. 26: 1031
360. Zinger B (1989) Synth. Met. 28: C37
361. Zinger B (1988) J. Electroanal. Chem. 244: 115
362. Tanguy J, Slama M, Hoclet M, Baudouin JL (1989) Synth. Met. 28: C145
363. Vieil E, Oudard JF, Servagent S (1988) Synth. Met. 28: C599
364. Yeu T, Nguen TV, White RE (1988) J. Electrochem. Soc. 135: 1971
365. Heinze J, Bilger R, Meerholz K (1988) Ber. Bunsenges. Phys. Chem. 92: 1266
366. Oudard JF, Allendoerfer RD, Osteryoung RA (1988) J. Electroanal. Chem. 241: 231
367. Feldberg SW, Rubinstein I (1988) J. Electroanal. Chem. 240: 1
368. Hillman AR, Mallen EF, Hammett HA (1988) J. Electroanal. Chem. 244: 353
369. Hamnett A, Hillman AR (1988) J. Electrochem. Soc. 135: 2517
370. Hillman AR, Mallen EF (1988) J. Electroanal. Chem. 243: 403
371. Cunningham DD, Galal A, Pham CV, Lewis ET, Burkhardt A, Laguren-Davidson L, Nkansah A, Ataman OY, Zimmer H, Mark Jr HB (1988) J. Electrochem. Soc. 135: 2750
372. Novak P, Inganas O (1988) J. Electrochem. Soc. 135: 2485
373. Janßen W, Beck F (1988) Polymer 30: 353
374. Dubois JC, Gagnes O (1989) Synth. Met. 28: C871
375. Hirai T, Kuwabata S, Yoneyama H (1988) J. Electrochem. Soc. 135: 1132
376. Kuwabata S, Ito S, Yoneyama H (1988) J. Electrochem. Soc. 135: 1691
377. Gningue D, Horowitz G, Garnier F (1988) J. Electrochem. Soc. 135: 1695
378. Maxfield M, Jow TR, Gould S, Sewchok MG, Shacklette LW (1988) J. Electrochem. Soc. 135: 299
379. Furukawa Y, Veda F, Hyodo Y, Harada I, Nakajama T, Kawagoe T (1988) Macromolecules 21: 1297
380. Tang X, Sun Y, Wei Y (1988) Makromol. Chem., Rapid Commun. 9: 829
381. Ni S, Wang L, Wang F (1988) Polymer Bull. 20: 311
382. LaCroix JC, Diaz AF (1988) J. Electrochem. Soc. 135: 1457
383. Kovacic P, Timberlake JW (1988) Polymer J. 20: 819
384. Zotti G, Cattarin S, Comisso N (1988) J. Electroanal Chem. 239: 387
385. Stilwell DE, Park SM (1988) J. Electrochem. Soc. 135: 2254, 2491, 2497
386. Rubinstein I, Sabatani E, Rishpon J (1988) J. Electrochem. Soc. 135: 3078
387. Shacklette LW, Wolf FF, Gould S, Baughman RH (1988) J. Chem. Phys. 88: 3955
388. Nakajima T, Kawagoe T (1989) Synth. Met. 28: C629
389. Mizumoto M, Namba M, Nishimara S, Miyadera H, Koseki M, Kobayashi Y (1988) Synth. Met. 28: C639
390. Genies E, Hany P, Sautier Ch (1989) Synth. Met. 28: C647
391. Osaka T, Naoi K, Ogano S (1988) J. Electrochem. Soc. 135: 1071
392. Arbizzani C, Mastragostino M, Panero S, Prosperi P, Scrosati B (1989) Synth. Met. 28: C663
393. Mammone RJ, Binder M (1988) J. Electrochem. Soc. 135: 1057
394. Nagotomo T, Omoto O (1988) J. Electrochem. Soc. 135: 2124
395. Thakur M (1988) Macromolecules 21: 661

Chemically Modified Electrodes

Andreas Merz

Institute of Organic Chemistry, University of Regensburg, Universitätsstr. 31, D-8400 Regensburg, FRG

1 Introduction and Overview 51

2 Manufacturing with Modified Electrodes 52
  2.1 Coating with Preformed Polymers 52
  2.2 Formation of Polymer Coatings Directly from Monomers 54
     2.2.1 Non-electrochemical Methods 54
     2.2.2 Electrochemically Initiated Polymerization 56
     2.2.3 Electrochemical Preparation of Conductive Polymers 56
  2.3 Inorganic Supporting Structures 58
     2.3.1 Ion Exchanging Materials 58
     2.3.2 Porous Supporting Structures 59
  2.4 Characterization of Modified Electrodes 60

3 Requirements for Mediator Catalysis at Modified Electrodes 61
  3.1 Classification of Mediator Catalyzed Reactions 61
     3.1.1 "Outer-sphere" or Simple Redox Catalysis 61
     3.1.2 "Inner-sphere" or Chemical Redox Catalysis 62
     3.1.3 Surface Confined Mediator Systems 62
  3.2 Theory and Electroanalytical Chemistry 63
     3.2.1 General Description of Models 63
     3.2.2 Classification of Cases 63
     3.2.3 Electroanalytical Proof of Redox Catalysis 66
     3.2.4 Practical Problems 66

4 Preparative Scale Electrochemistry at Modified Electrodes 66
  4.1 General Overview 66
  4.2 Reduction of Vicinal Organic Dihalides 67
  4.3 Electrocatalytic Hydrogenation 69
  4.4 Vitamine B_{12} Reactions 69
  4.5 Ceramic Oxide Electrodes 71
  4.6 Modified Electrodes for Stereoselectivity of Electroorganic Synthesis . . 72

5 Analytical Applications 75
  5.1 Preconcentration Methods 75
  5.2 Microstructured Electrodes 76
      5.2.1 Multilayered Electrodes 77
      5.2.2 Microelectrode Arrays 77
      5.2.3 Sensors Based on Microelectrode Arrays 78
      5.2.4 Solid State Electrochemistry 79
  5.3 Reference Electrodes 80

6 Outlook 81

7 Note Added in Proof 82

8 References 83

The historical development of chemically electrodes is briefly outlined. Following recent trends, the manufacturing of modified electrodes is reviewed with emphasis on the more recent methods of electrochemical polymerization and on new ion exchanging materials. Surface derivatized electrodes are not treated in detail. The catalysis of electrochemical reactions is treated from the view of theory and of practical application. Promising experimental results are given in detail. Finally, recent advances of chemically modified electrodes in sensor techniques and in the construction of molecular electronics are given.

1 Introduction and Overview

In 1975, the fabrication of a *chiral electrode* by permanent attachment of amino acid residues to pendant groups on a graphite surface was reported [1]. At the same time, stimulated by the development of bonded phases on silica and aluminia surfaces [2] the first example of derivatized metal surfaces for use as *chemically modified electrodes* [3,4] was presented. A silanization technique was used for covalently binding redox species to hydroxy groups of SnO_2 or Pt surfaces. Before that time, some successful attemps to create electrode surfaces with deliberate chemical properties made use of specific adsorption techniques [5].

These publications stimulated innovative work in many research groups as documented in the first major account on the subject in 1980 [6]. By this time it was also recognized that the modification of electrode surfaces by polymer layers can be simpler, more effective and more versatile than the derivatization techniques [7,8,9].

From the beginning, most electrode coatings were designed to contain electroactive groups. The primary advantage was that such coatings can be easily monitored and characterized by electroanalytical methods in situ whereas other physical methods to characterize solid-liquid interfaces are still not well developed. The most fascinating and actually stimulating idea about surface confined redox molecules, however, is the possibility of controlling surface properties by electrode potentials. Thus a number of well recognized fields of applications have been developed over the last 10 years. This is well documented in R. W. Murray's thorough and authoritative review of 1984 [10]:

— study of electron self exchange within surface films
— potential dependent color display
— corrosion protection and photoactivation of photovoltaic electrodes
— catalysis of electrochemical reactions
— reference electrode systems
— substrate selective electroanalytical sensors
— enzyme immobilization to electrodes for biological sensors
— molecular based electronic devices

Provided electron transfer between the electrode and solute species is not interrupted by the coating, even electroinactive films can offer interesting applications. Thus, a chiral environment in the surface layer may impose stereoselectivity in the follow-up reactions of organic or organometallic intermediates. Furthermore, polymer layers may be used to obtain diffusional permeation selectivity for certain substrates, or as a preconcentration medium for analyzing low concentration species.

In the first part of the present review, new techniques of preparation of modified electrodes and their electrochemical properties are presented. The second part is devoted to applications based on electrochemical reactions of solute species at modified electrodes. Special focus is given to the general requirements for the use of modified electrodes in synthetic and analytical organic electrochemistry. The subject has been reviewed several times [11-14]. Besides the latest general review by Murray [10], a number of more recent overview articles have specialized on certain aspects: macromolecular electronics [15-17], theoretical aspects of electrocatalysis [18], organic applications [19-21], sensor electrodes [22], and applications in biological [23,24] and medicinal [25] chemistry.

Andreas Merz

2 Manufacturing of Modified Electrodes

2.1 Coating with Preformed Polymers

Surface derivatization methods [6] cannot produce more than monolayer ($\leq 5 \times 10^{-10}$ mol cm^{-2}) coverages of material at the electrode surface. Whenever this is sufficient, the use of surface functionalities like carboxyl groups on carbon materials [1] or hydroxy groups on metal surfaces [3,4,16] is convenient for the attachment of an unlimited variety of molecular species. In most cases of application however, multilayers of active surface coverage are required. This is best achieved with polymer layers and in the following this review will be dealing almost exclusively with organic polymer coated electrodes and some inorganic equivalents.

It was first shown that electrochemistry, at reasonable electron transfer rates, is possible at electrodes coated with substantial amounts of *electroinactive* polymer [26]. The first polymer coatings on electrodes containing redox centers, polyvinylferrocene or polynitrostyrene, were reported soon after [7,8]. The most important result of this early work was the fact that many more redox centers than monolayer populations are reversibly addressed by electrochemical reduction/oxidation and that again the electrochemistry of solute species can be observed at such polymer modified electrodes. Cyclic voltammograms of e.g. diphenylanthracene at bare or polyvinylferrocene-coated platinum electrodes are virtually identical [7,27]. The presence of pinholes in the polymer film was discussed to explain the fact that solute species could penetrate the film so easily [7], but the present view is that substrates and reactants are transported through the polymer film by diffusion in the splvent-swollen polymer [27]. Even with relatively small diffusion coefficients of species in the film, voltammograms look the same on coated and uncoated electrodes when the diffusion layer in relation to the time scale of the experiment is large compared to film thickness and pore size [28]. Relatively free diffusion appears to be characteristic of linear vinylic polymers and, as will be shown later, is quite essential for applications in electrocatalysis. The diffusion coefficients can be estimated from the comparison of limiting currents of rotating disk voltammograms at coated and uncoated electrodes [29,30]. Values in the range of 10^{-10} cm^2 s^{-1} may be typical.

The techniques of applying preformed polymers to the surface require that the polymer is soluble in some solvent and insoluble in the solvent/supporting electrolyte system used for electroanalytical studies. Simple "dip coating" and subsequent drying is effective in producing the films although "wiping away" [7] or "shaking away" [26] excess dipping solution before evaporation of the adherent solvent point to a certain lack of reproducibility. The nature of adhesion of the films to the electrode surface is not known for certain and has been described as adsorption or simply precipitation. A confirmed chemical binding of the polymer layer can be achieved by the use of surface attached anchor groups [31,32]. In the case of polyvinylferrocene, enhanced adsorption in the oxidized state allows an electrochemical procedure to control the polymer layer formation [7].

Alternatively, the amount of polymer deposited can be determined via "droplet evaporation" by microsyringe application of known amounts of polymer solution on the electrode surface, but the problem of homogeneity of the film is still not solved.

The reproducibility of polymer film formation is greatly improved by the "spin coating" technique where the polymer solution is applied by a microsyringe onto the center of a rapidly rotated disk electrode [33]. Rather thick films can be produced by repeated application of small volumes of stock solution. A thorough discussion and detailed experimental description of a reliable spin coating procedure was given recently [34].

A great variety of suitable polymers is accessible by polymerization of vinylic monomers, or by reaction of alcohols or amines with functionalized polymers such as chloromethylated polystyrene or methacryloylchloride. The functionality in the polymer may also be a ligand which can bind transition metal complexes. Examples are poly-4-vinylpyridine [29, 35–40] and triphenylphosphine modified polymers [41]. In all cases of reactively functionalized polymers, the loading with redox active species may also occur after film formation on the electrode surface but it was recognized that such a procedure may lead to inhomogeneous distribution of redox centers in the film [36].

A very similar idea is the use of ion exchanging polymers into which redox active ions can be incorporated by equilibration with an adherent solution. Sulfonated perfluoro polymers (Nafion), [42–48], polyvinylsulfonic acid [49], and polystyrene sulfonic acids [30, 34, 50, 51] have been introduced as cation exchanging films, and protonated or quaternized aminopolymers such as polylysine [52, 53] and protonated polyvinylpyridine [54] or N-methylated polyvinylpyridine [55] were used as anion exchangers. $[Fe(CN)_6]^{4-}$ [54] and $[Co(C_2O_4)_3]^{3-}$ [52, 53], for instance, were used as redox active anions and 2,2'-bipyridine complexes of Ru [43, 45, 50, 51] and Co [46, 47], $Fe^{II}(edta)^-$, and viologen cations [48] were used as loading ions for cation exchanging polymers. The ion exchanging process of introducing redox active ions may be enforced by cycling the potential of the polymer coated electrode in contact with the feeding solution [30].

Polyelectrolytes may exhibit rather specific properties depending on their morphology. Nafion, for instance, contains extremely hydrophobic fluorocarbon regions where neutral organic species [44, 56, 57] may be selectively incorporated. Other polyelectrolytes appear to have domains filled with solvent where ions can diffuse quite freely [52]. In such polymers, charge may propagate by both electron hopping and diffusion of redox sites [53]. Although polyelectrolytes offer some principal advantages over covalent polymers, a number of draw-backs have been identified in some polymers: lack of strong binding to the electrode surface, insufficient ion exchange capacity, insufficient retention of redox ions during measurements [58], relatively high solubility in aqueous solvents, and almost complete impenetrability in nonaqueous solvents unless hydrated [51]. New ion exchanging copolymers were recently reported

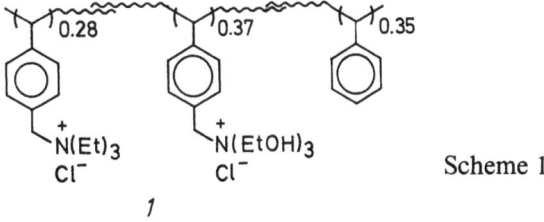

Scheme 1

that apparently overcome most, or all of these difficulties. A ternary random copolymer *1* made from a polystyrene/chloromethylated styrene copolymer by subsequent quaternization with two different tertiary amines at the chloromethyl sites exhibits enhanced diffusion rates of ions within the film and very good retention (Scheme 1 [58, 59]). Polystyrene sulfonic acids can be stabilized for use in aqueous solutions by copolymerizing with polyvinylpyridine [30] and crosslinking of the polymer after filming with 1,2-dibromoethane vapors [60]. Both the water solubility and the problems with ion retention may be avoided by the use of mixed polyanions and polycations such as Nafion or polystyrenesulfonate and polyviologenes. Polysalts are precipitated on electrode surfaces by simply mixing aqueous solutions of both components [61,62].

2.2 Formation of Polymer Coatings Directly from Monomers

2.2.1 Non-electrochemical Methods

Some chemically stable vinyl monomers can be polymerized from the gas phase using a glow discharge plasma for activation [63]. The polymer is produced as a tightly adhering film on surfaces. This method was applied in the preparation of polyvinyl ferrocene films from the monomer. The electrochemical properties of the films are similar to those made from soluble linear polyvinylferrocene [64,65]. Recent studies carried out in great detail show that the plasma polymerized PVF is more difficult to penetrate for electrolyte anions [66,67,68]. A rather high degree of crosslinking is

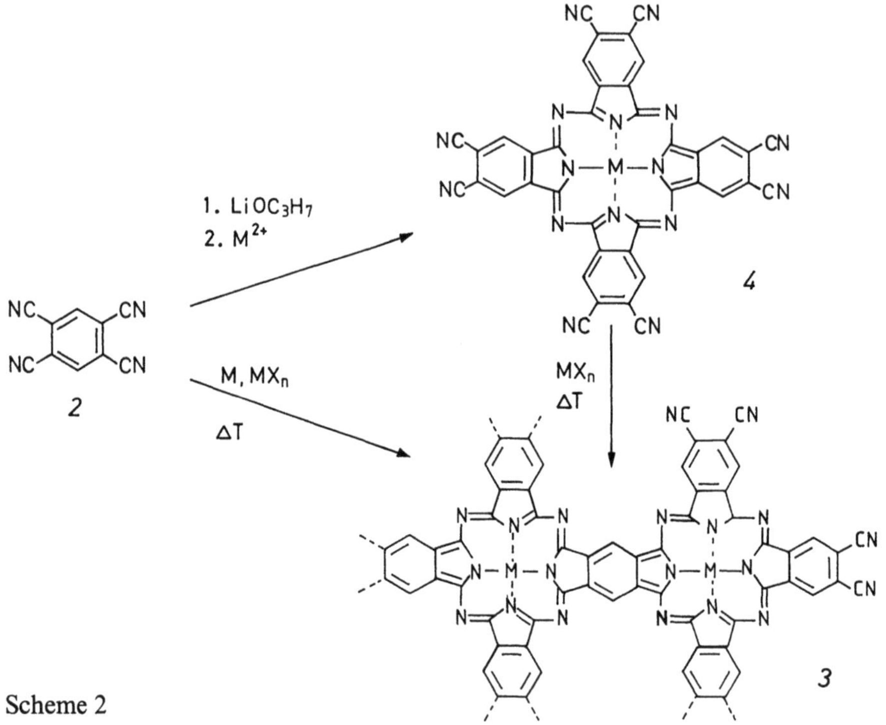

Scheme 2

believed to be responsible. Likewise, plasma polymerized polyvinyl pyridine has a complexing behaviour towards transition metal cations that is very different from that observed with the linear polymer [69]. On the whole, this method appears to be restricted to rather stable monomers and of limited value for electrode coatings that require a high permeability for electrolyte ions and substrates.

Polymeric phthalocyanines 3 are available from 1,2,4,5-tetracyanobenzene 2 by deposition from the vapor phase on hot substrate surfaces, or by thermal curing of its tetramer, octacyanophthalocyanine 4 in the presence of metal cations (Scheme 2 [70-74]).

On the whole, curing procedures appear a promising way to obtain very stable polymer films. Thus, the structure of already mentioned polylysine [52,53] has been revised as a block polymer involving either the α or ε amino groups of lysine [75]. Vitamin B_{12} modified carbon electrodes were prepared by thermal curing of a mixture of a diamino functionalized B_{12} derivative 5 and an epoxy prepolymer 6 of the araldite type (Scheme 3 [76]).

This process is probably accompanied by fixation at surface functional groups as well as crosslinking reactions. The simplicity of this approach makes it quite promising for a more general application.

The coupling of molecular species to surfaces by the silanization technique has been extended by employing bis-trialkoxysilylated monomers. Examples are the viologen derivative 7 [77,78] and the cobaltocinium complex 8 (Scheme 4 [79]). In acetonitrile containing trace amounts of water, siloxane type connections lead to simultaneous attachment to surface groups and crosslinking between redox centers.

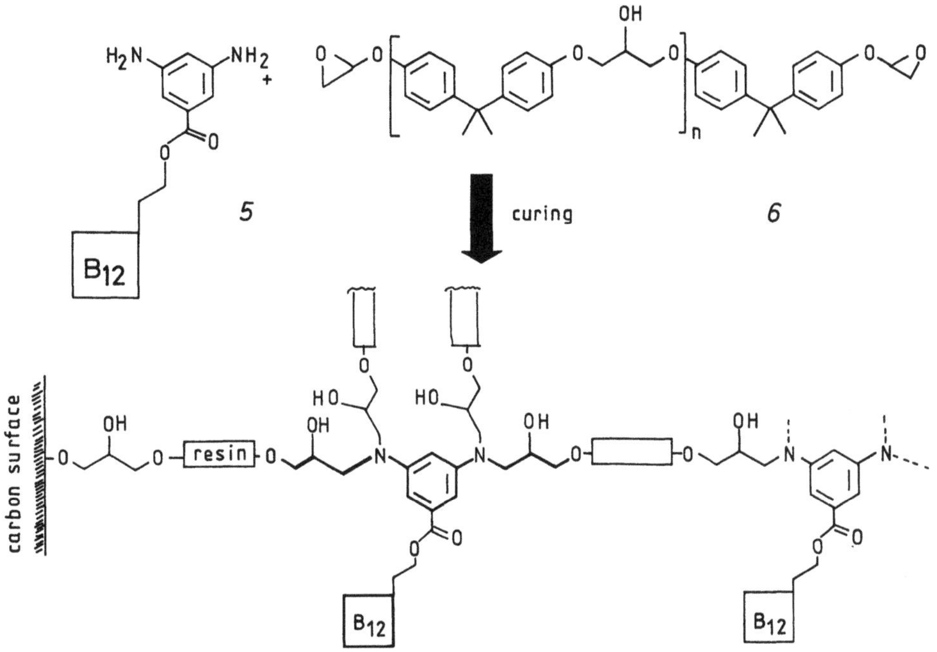

Scheme 3

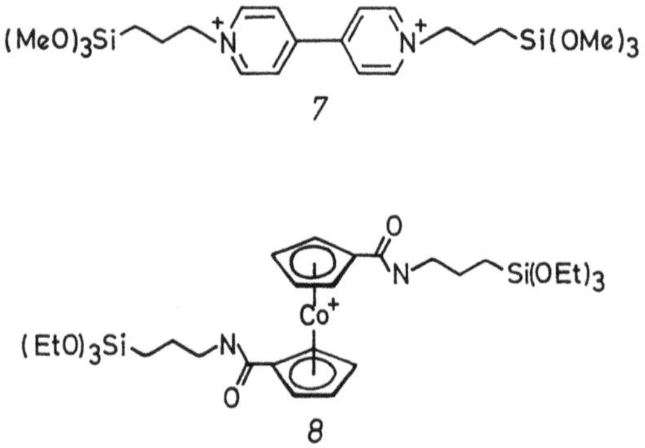

Scheme 4

2.2.2 Electrochemically Initiated Polymerization

Reactive radical ions, cations and anions are frequent intermediates in organic electrode reactions and they can serve as polymerization initiators, e.g. for vinylic polymerization. The idea of electrochemically induced polymerization of monomers has been occasionally pursued and the principle has in fact been demonstrated for a number of polymers [80,81]. But it appears that apart from special cases with anionic initiation the heterogeneous initiation is unfavorable and thus not competitive for the production of bulk polymers [81]. A further adverse effect is the coating of electrodes with insulating polymer layers, although the formation of insulating coatings can have advantages in the field of corrosion protection [82,83] and the electrochemical deposition of paint [84].

Electropolymerization is also an attractive method for the preparation of modified electrodes. In this case it is necessary that the forming film is conductive or permeable for supporting electrolyte and substrates. Film formation of nonelectroactive polymers can proceed until diffusion of electroactive species to the electrode surface becomes negligible. Thus, a variety of nonconducting thin films have been obtained by electrochemical oxidation of aromatic phenols and amines [85,86]. Some of these polymers have ligand properties and can be made electroactive by subsequent incorporation of transition metal ions [87–89].

Vinyl substituted bipyridine complexes [90] of ruthenium 9 and osmium 10 can be electropolymerized directly onto electrode surfaces [91,92]. The polymerization is initiated and controlled by stepping or cycling the electrode potential between positive and negative values and it is more successful when the number of vinyl groups in the complexes is increased, as in 11 [92]. A series of new vinyl substituted terpyridinyl ligands have recently been synthesized whose iron, cobalt and ruthenium complexes 12 are also susceptible to electropolymerization [93].

2.2.3 Electrochemical Preparation of Conductive Polymers

In 1979, the formation of conductive polypyrrole films by the electrochemical oxidation of pyrrole was reported for the first time [94]. This work has stimulated intense and fruitful research in the field of organic conducting polymers. Further important conductive polymers are polythiophene, polyaniline and polyparaphenylene. The development and technological aspects of this expanding research area is covered

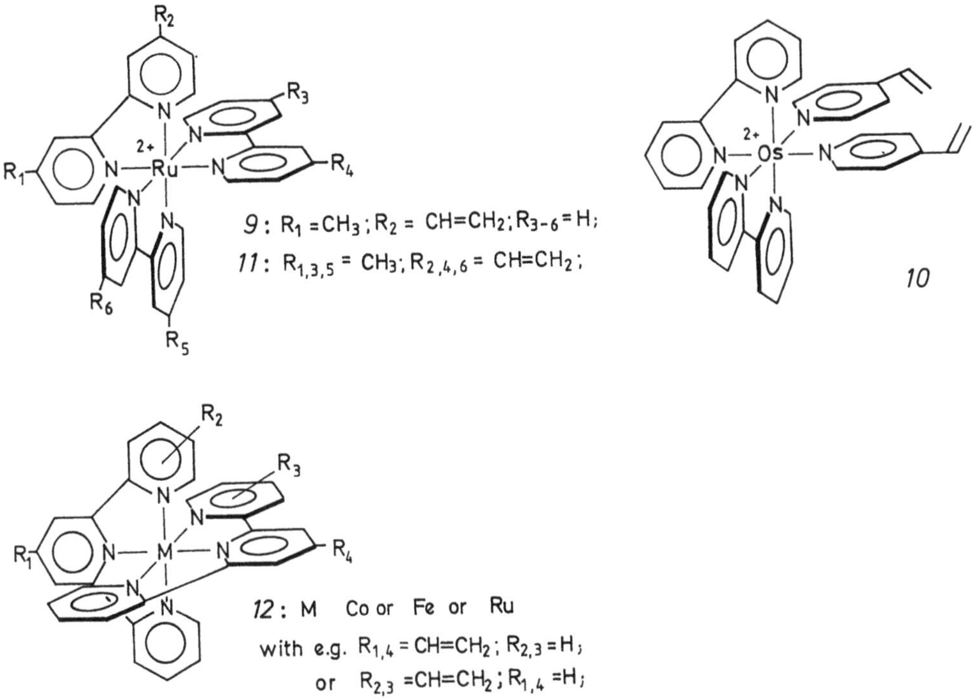

9: $R_1 = CH_3$; $R_2 = CH=CH_2$; $R_{3-6} = H$;
11: $R_{1,3,5} = CH_3$; $R_{2,4,6} = CH=CH_2$;

10

12: M Co or Fe or Ru
with e.g. $R_{1,4} = CH=CH_2$; $R_{2,3} = H$;
or $R_{2,3} = CH=CH_2$; $R_{1,4} = H$;

Scheme 5

in an excellent handbook [95] and the electrochemical aspects are reviewed in this volume [96] with respect to the mechanisms of formation of conducting polymers, the origin of their conductivity and their electroanalytical chemistry. Here, the properties of polypyrrole as a new type of electrode material or as a basic constituent of modified electrodes are discussed.

Due to its electronic conductivity, polypyrrole can be grown to considerable thickness. It also constitutes, by itself, as a film on platinum or gold, a new type of electrode surface that exhibits catalytic activity in the electrochemical oxidation of ascorbic acid and dopamine [97], in the reversible redox reactions of hydroquinones [98,99] and the reduction of molecular oxygen [100,101]. N-substituted pyrroles are excellent electropolymerizable monomers for the preparation of conductive polymers containing additional redox centers. Several research groups have used this approach to obtain modified electrodes. The polymers obtained in this way are summarized in Table 1.

Even the simplest N-substituted pyrrole, N-methylpyrrole, electropolymerizes much slower than pyrrole itself [117,118]. Furthermore, the conductivity of polymers derived from N-substituted pyrroles is much lower than that of pure polypyrrole [105,118], apparently due to the non-planarity of polypyrrole chains induced by the bulky N-substituents. The very low conductivity of higher poly-N-alkylpyrroles [118] as well as most homopolymers given in Table 1 allows only for the formation of rather thin films. Nevertheless, such films have proved valuable as modified electrodes as will be detailed in Sect. 4.

The homo-electropolymerization is improved by the use of monomers with more

Table 1. N-substituted pyrroles employed in the preparation of redox-Modified polypyrrole films on electrodes

Active group	Homopolymer	Co-Polymer
Aryl groups	102)	103, 104, 105)
Ferrocenes	103)	106, 107)
Nitroxides	108)	109)
Pyridine (ligand)	110)	
Bipyridine (ligand)	111, 112)	109)
Triarylamine	109)	109)
Viologene	113)	
Anthraquinone	114)	
Metalloporphyrines	115, 116, 295–297)	

than one pyrrole nucleus connected to the active redox group [108]. Another approach is the technique of co-polymerization of N-substituted pyrroles with the readily polymerizable unsubstituted pyrrole [106]. In this case, however, the polymerization kinetics and polymer composition as well as the conductivities are not a linear function of the monomer ratio in the feeding solution [107]. Thus, in the preparation of poly-[pyrrole/N-(ω-ferrocenylalkyl-)pyrrole] a molar fraction of 0.1 of the substituted pyrrole in the polymer could not be exceeded when the typical properties of polypyrrole were to be maintained. Within this limit, self-supported modified polypyrrole films are also available [107, 119].

The permanent inclusion of solution constituents during the electropolymerization of pyrrole was also used to prepare modified electrodes with cobalt phthalocyanine [120, 121] and glucose oxidase [122, 123] as examples. Another technique of perhaps more general application is the preparation of composite films [124–128] where pyrrole is electropolymerized into a preformed solvent swollen film of a redox polymer at an electrode [129, 130]. A great variety may be expected when combinations of dozens of redox polymers with at least 6 known conductive organic polymers obtainable by anodic polymerization are taken into account. As will be shown in Sect. 3, it will be necessary to produce conductive modified polymers with a high degree of porosity, with pore diameters at least one order of magnitude larger than molecular dimensions. Particularly for this goal, composite polymers involving conductive components appear promising.

2.3 Inorganic Supporting Structures

2.3.1 Ion Exchanging Materials

Polynuclear transition metal cyanides such as the well-known Prussian blue and its analogues with osmium and ruthenium have been intensely studied [131]. Prussian blue films on electrodes are formed as microcrystalline materials by the electrochemical reduction of $FeFe(CN)_6$ in aqueous solution [132, 133]. They show two reversible redox reactions, and due to the intense color of the single oxidation states, they appear to be candidates for electrochromic displays [134]. Ion exchange properties in the reduced state are limited to certain ions having similar ionic radii. Thus, the reversible electrochemistry is retained in the presence of K, Rb, Cs, and NH_3 cations but blocked by smaller cations [135]. An application as ion-selective electrodes has therefore been discussed [131].

The clay mineral bentonite (sodium montmorillonite) has an excellent ion exchange and adsorption capacity. Films can be applied to electrode surfaces from colloidal clay solutions by simple dip or spin coating that become electroactive after incorporation of electroactive cations or metal particles [136-143].

Clay films cast from a pure aqueous colloid appear to form a regular array of microplatelets, thin films of which show selective cation exchange, e.g. segregation of $Ru(bipy)_3^{2+}$ from Na^+ and methylviologen dication [138], and even partial separation of the enantiomers of $Co(bipy)_3$ [143]. Thicker films (approx. 3 µm) can be supported by the addition of polyvinyl alcohol [136,138]. This additive also aids swelling of the bentonite structure giving rise to more ready ion exchange and mobility of cations [138]. Clay minerals have recently been introduced as supporting structures for redox reagents in organic reactions [144]. Since they are inexpensive and have excellent chemical stability they may be promising for applications in organic electrochemistry.

Similar films are obtained from powdered molecular sieves loaded with organic molecules [145]. Zeolite Y microparticles embedded into a polystyrene film and loaded with appropriately sized transition metal complexes allow selective electron exchange reactions between trapped and mobile species in the film [146].

2.3.2 Porous Supporting Structures

Some porous ceramic structures of oxides on titanium (Cr_2O_3, RuO_2, MnO_2, VO_x) obtained by baking films of metal complexes like acetylacetonates on titanium surfaces can also be regarded as chemically modified electrodes [147-149]. Applications in organic electrochemistry will be discussed in Sect. 4.

A specially prepared alumina has been suggested as a support for organic polymers [150]: the anodic oxidation of aluminum surfaces in dilute phosphoric acid gives rise to the formation of an oxide layer with parallel pores perpendicular to the metal surface. The size distribution of the pores is rather narrow and can be controlled (20 to 150 nm) by the cell voltage. The porous oxide layer can then be cleaved off the aluminum surface by amalgamation of the metallic site and subsequently be transferred to metal surfaces such as gold. A very stable redox electrode has been described, consisting of a polyvinyl pyridine film coating just the inner walls of the pores.

Yet another novel approach towards inorganic supports is the use of microporous glass films prepared on metal or carbon surfaces from glass sols by spin coating and subsequent thermal treatment [151]. The enhanced mechanical stability of such devices has been pointed out. Experimental results available up to now show that the characteristic electrochemistry of a platinum oxide surface in aqueous sulfuric acid is still observed after glass coating, that the diffusional barrier of the glass films can be manipulated by manufacturing conditions, and that electroactive polymers can be formed within the glass layer that adhere very tightly unless they are protruding over the glass surface.

2.4 Characterization of Modified Electrodes

For the in situ characterization of modified electrodes, the method of choice is electrochemical analysis by cyclic voltammetry, ac voltammetry, chronoamperometry or chronocoulometry, or rotating disk voltametry. Cyclic voltammograms are easy to interpret from a qualitative point of view (Fig. 1). The other methods are less direct but they can yield quantitative data more readily.

Further structural information is available from physical methods of surface analysis [152] such as scanning electron microscopy (SEM), X-ray photoelectron or Auger electron spectroscopy (XPS), or secondary-ion mass spectrometry (SIMS), and transmission or reflectance IR and UV/VIS spectroscopy. The application of both electroanalytical and surface spectroscopic methods has been thoroughly reviewed [10] and appropriate methods are given in most of the references of this chapter. The physical methods mostly require ultra high vacuum conditions having the disadvantage of not being applicable directly to solvent swollen films, but recent developments of in situ measurements in SIMS [153], X-ray diffraction [154], surface enhanced Raman spectroscopy (SERS) [155], and scanning electrochemical tunneling microscopy (SETM) [156] give hope that the full structural characterization of modified electrodes will be achieved in the near future.

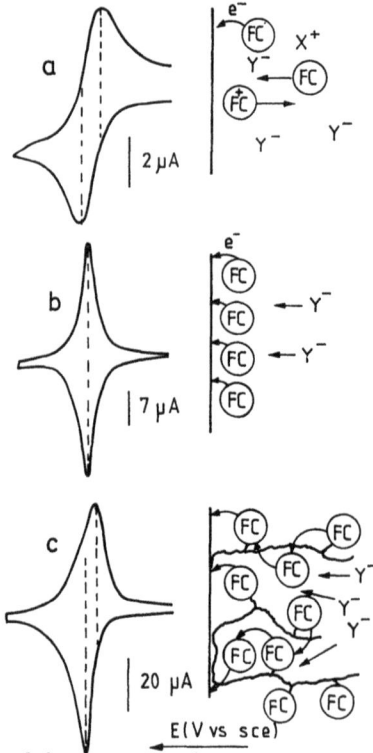

Fig. 1a–c. Cyclic voltammograms of dissolved and surface confined ferrocene in acetonitrile/0.1 M TBAP. a. 4×10^{-4} M dissolved ferrocene at Pt (ref. [6]). b. 4-ferrocenyl-phenylacetamid monolayer bound to Pt (ref. [6]). c. Polyvinylferrocene dip coated on Pt, $\Gamma = 1 \times 10^{-9}$ mol cm^{-1}. Straight arrows indicate diffusional events. Curved arrows electron transfer events (from ref. [9]).

3 Requirements for Mediator Catalysis at Modified Electrodes

3.1 Classification of Mediator Catalyzed Reactions

Mediated electrolyses make use of electron transfer mediators P/Q that shuttle electrons between electrodes and substrates S, avoiding adverse effects encountered with the direct heterogeneous reaction of substrates at electrode surfaces (Scheme 6). In recent years this mode of electrochemical synthesis has been widely studied and it is becoming increasingly well understood. A review is given in vol 1 of the present electrochemistry series [157].

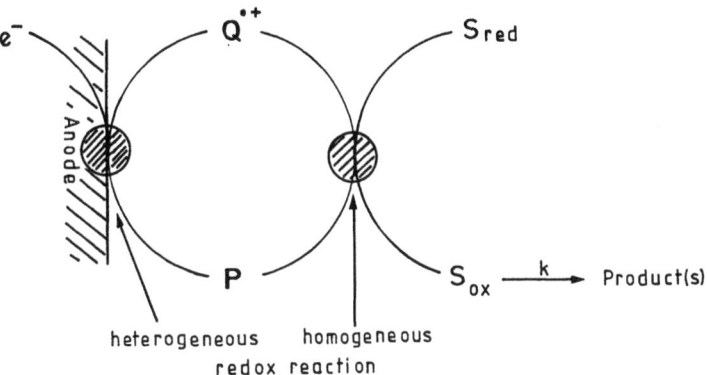

Scheme 6

If the redox mediator is dissolved in the electrolyte solution together with the substrate it is a *homogeneous* mediator as opposed to the surface bound species at modified electrodes. Two basic cases of homogeneous mediation have been classified [158]:

3.1.1 "Outer sphere" or Simple Redox Catalysis

The oxidation or reduction of a substrate suffering from sluggish electron transfer kinetics at the electrode surface is mediated by a redox system that can exchange electrons rapidly with the electrode and the substrate. The situation is clear when the half-wave potential of the mediator is equal to or more positive than that of the substrate (for oxidations, and vice versa for reductions). The mediated reaction path is favored over direct electrochemistry of the substrate at the electrode because, by the diffusion/reaction layer of the redox mediator, the electron transfer step takes place in a three-dimensional reaction zone rather than at the surface [158].

Mediation can also occur when the half-wave potential of the mediator is on the thermodynamically less favorable side, in cases where the redox equilibrium between mediator and substrate is disturbed by an irreversible follow-up reaction of the latter. The requirement of sufficiently fast electron transfer reactions of the mediator is usually fulfilled by such reversible redox couples P/Q in which bond and solvate reorganization energies are small due to enhanced delocalization of charge. Typical examples are organic molecules with extended π-systems, and transition metal complexes many of which are given in Ref. [157] together with further requirements for useful mediators.

3.1.2 "Inner-sphere" or Chemical Redox Catalysis

When the overall electrochemical reaction of the substrate involves bond-breaking or bond-forming processes prior to, followed by, or concerted with electron transfer, or between successive electron transfer steps, the overall activation energy may be lowered by "chemical" redox catalysis. In this situation the catalyst has to play a multiple role in the reaction path: electron tranfer *and* intermediate fixation of substrates and reactive intermediates. Substrate and product selectivity will arise herefrom. One remarkable feature of chemical redox catalysis is the possibility of concerted two electron transfer reactions which is generally not possible with outer-sphere redox reactions [159]. Clearly, the catalyst will have to be tailormade to meet the electronic and structural requirements of the substrate, and the most serious problem is to regenerate such catalysts in their active forms [159,160].

Models for this goal are the reactions of redox enzymes which in turn should be closest possibly mimicked by transition metal complex chemistry. A presently very promising approach is therefore the use of enzymes or coenzymes for the desired reaction in combination with a redox mediator that regenerates the redox activity of the enzyme complex [23,24,159]. Such a system would be called a multiple mediator arrangement.

3.1.3 Surface Confined Mediator Systems

Electrocatalysis with mediators located in coatings at the electrode surface is one of the proposed applications of modified electrodes. A number of obvious advantages over the homogeneous case can easily be compiled:
— Though using a much lower total amount of catalyst than in the homogeneous case, a considerably higher catalyst concentration in the reaction layer can be supplied.
— Theory predicts a substantial rate enhancement over the homogeneous case.
— Even the use of costly catalysts may be considered
— No separation of products and catalyst is required
 A number of possible pitfalls can be anticipated as well:
— A catalyst molecule bound to a polymeric support may be less active than in homogeneous solution, or exhibit different reactivity [161].
— When a small fraction of irreversible mediator side reactions cause a rapid decrease of catalytic activity in the homogeneous case [157], in a modified electrode this would be disastrous since there is no bulk supply of catalyst. Thus, higher turnover numbers are generally required than in the homogeneous case
— Reactions of reactive intermediates with the polymer constituents may produce unforeseeable artefacts.
— Polymers or other supporting materials may not be able to tolerate the high current densities desired for preparative electrolyses.

3.2 Theory and Electroanalytical Chemistry

3.2.1 General Description of Models

A general theory based on the quantitative treatment of the reaction layer profile exists for pure redox catalysis where the crucial function of the redox mediator is solely electron transfer and where the catalytic activity largely depends only on the redox potential and not on the structure of the catalyst [158, 162]. This theory is consistent with experimental data and has been successfully applied in the evaluation of kinetic and thermodynamic parameters of electrochemical reactions [163–166].

This theoretical model has been extended to the case of heterogeneous mediation at redox modified electrodes [167–174]. Since the generation of a three-dimensional reaction/diffusion layer is essential for redox catalysis to occur, a monolayer derivatized electrode is expected to be inactive in the pure redox catalysis case. A polymer layer containing mutilayer equivalents of redox centers thus appears to be ideally suited, and at first sight, the film may be expected to operate better the thicker it is [167]. But even a qualitative view can show that, apart from the electron transfer rate of the actual mediation step, two further potentially rate determining processes must be considered: the propagation of charge within or across the film, and the partition and mobility of substrates (and products) in the film as compared to the bulk solution.

Charge propagation within the film is in principle slower than charge injection (or consumption) at the electrode/film interface [175]. Whether electrons are transported through the film by electron hopping between fixed redox sites [176], or partially by diffusion of redox sites within the film [53], or assisted by movements of the polymer chains [91], the overall process is treated as a diffusion of charge. The corresponding diffusion coefficient is dependent on the intrinsic self-exchange rate and on the concentration of redox sites in the film [36, 177, 178]. The diffusion coefficients of substrates in the film may well be 3 or 4 magnitudes smaller than in a solvent [29, 30].

If substrate diffusion becomes rate determining, only a small fraction of the film at the film/solution interface will be used. On the other hand, if charge diffusion becomes rate determining, the catalytic reaction can take place only in a film fraction close to the electrode surface. Each of these effects will render parts of the film superfluous, and it is obvious that there is no sense in designing very thick redox films, rather there is an optimal layer thickness to be expected depending on the individual system.

3.2.2 Classification of Cases

Rigorous quantitative treatments lead to kinetic zone diagrams that distinguish between different extreme and borderline cases depending on which parameters controll the overall reaction [170, 172] (Fig. 2). A realistic picture is obtained from cross sectional diagrams of the electrode-film-solution interface that show the concentration profiles of catalyst redox states and substrate for the various kinetic situations under steady state operation (Fig. 3 [18, 168, 174]). The catalytic efficiency can be expressed by a characteristic current [170] as given in Fig. 2 for the limiting current at the rotating disk electrode, or by an effective heterogeneous rate constant [172].

Apparently, for effective catalysis, conditions are desirable under which the complete film volume is active (case R, Fig. 3). But such conditions will be difficult to

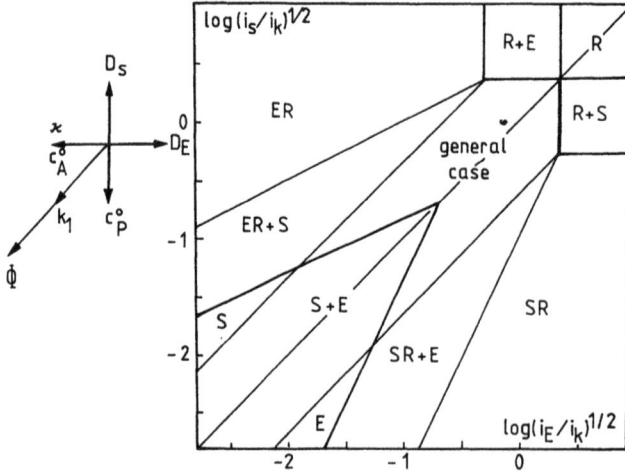

Fig. 2a–c. Kinetic zone diagram for the catalysis at redox modified electrodes[a,b,c]. **a.** The kinetic zones are characterized by capital letters: R: control by rate of mediation reaction, S: control by rate of subtrate diffusion, E: control by electron diffusion rate, combinations are mixed and borderline cases; **b.** The kinetic parameters on the axes are given in the form of characteristic currents: i_k: current due to exchange reaction, i_E: current due to electron diffusion, i_S: current due to substrate diffusion; **c.** The signpost on the left indicates how a position in the diagram will move on changing experimental parameters: c_A^0: bulk concentration of substrate; c_P^0, c_Q^0: catalyst concentration in the film; D_S, D_E: diffusion coefficients of substrate and electrons; k_1: rate constant of exchange reaction; k: distribution coefficient of substrate between film and solution; Φ: film thickness (from ref. [174]).

Table 2. Analytical scale mediated redox reactions with participation of a significant layer of polymer films

Polymer-System	Mediator	Reaction Catalyzed	Ref.
Polyvinylpyridine/H$^+$	IrCl$_6^{2-}$	Fe^{2+}/Fe^{3+}	181)
Polyvinylpyridine/H$^+$	Ru^{II}(edta)	[Fe(CN)$_6$]$^{4-/3-}$	181)
Polylysine/H$^+$	[Mo(CN)$_8$]$^{3-}$	Co(tpy)$_2^{2+}$	175)
Polylysine/H$^+$	[W(CN)$_8$]$^{3-}$	Co(tpy)$_2^{2+}$	175)
Polyvinypyridine/Polylysine/H$^+$	IrCl$_6^{2-}$	Fe^{2+}/Fe^{3+}	182)
Nafion	Methylviologen or Polyxylylviologen	O$_2$ redn.	183)
Nafion	[Ru(NH$_3$)$_6$]$^{2+}$/ Co^{II}TPMPyP	O$_2$ redn.	180)
Polyvinylpyridine-ligand	Ru^{II}(bipy)$_2$Cl →	Fe^{2+}/Fe^{2+}	29)
Polyvinylpyridine/Polystyrenesulfonate crosslinked	Os(bipy)$_3^{2+}$	[Fe(CN)$_6$]$^{4-/3-}$	30)

Abbreviations: edta = ethylenediamine-tetraacetic acid, bipy = 2,2'bipyridine, tpy = 2,2',6',2''-terpyridine, TPMPyP = tetrakis(4-N-methylpyridyl)porphyrin

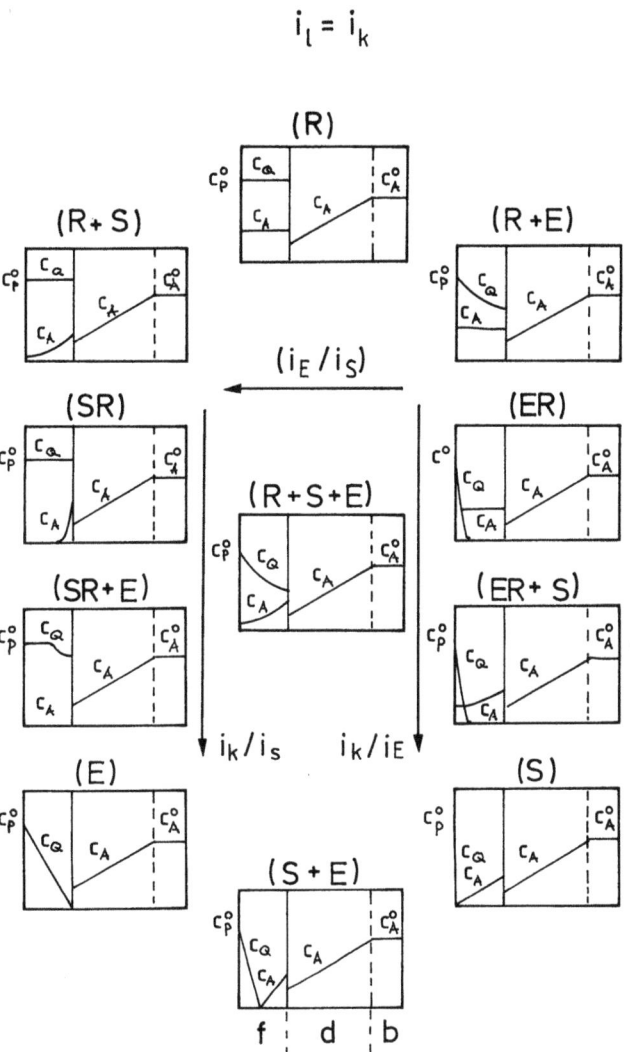

Fig. 3. Steady state concentration profiles of catalyst and substrate species in the film and diffusion layer for for various cases of redox catalysis at polymer-modified electrodes. Explanation of layers see bottom case (S + E): f: film; d: diffusion layer; b: bulk solution; i_l: limiting current at the rotating disk electrode; other symbols have the same meaning as in Fig. 2 (from ref. [168]).

achieve in practice. The second-best situation is a reaction layer extending from the film/solution interface as far as possible towards the electrode surface (case R + S, Fig. 3). If such a situation can be established, one can calculate that electrosynthesis at current densities up to 1 A cm^{-2} are possible with a rate enhancement up to 10^3 over the uncatalyzed electrode reaction [172]. The necessary but not easily available set of parameters includes a second order rate constant for the mediation step > 10 mol^{-1} L^{-1}, and a diffusion coefficient of the substrate in the layer near 10^{-6} cm^2 × s^{-1}. The optimum film layer under these conditions is 1 µm [172]. All other cases in the kinetic zone diagram should be less or not effective with the exception of *chemical catalysis* by the mediator [76,179,180]. In this latter case even a catalyst monolayer would be catalytically active if a sufficient turnover number is given.

3.2.3 Electroanalytical Proof of Redox Catalysis

Appropriate electroanalytical procedures to verify the one or other case have been given in the references of this section. The main techniques are cyclic voltammetry, chronoamperometry, chronocoulometry, and rotating disk voltammetry. The last one appears to be best suited since constant mass transport in the film is a very important feature as outlined above [18,29,30,170,172,173]. Table 2 gives examples for which the participation of a substantial part of the film (R or S + R cases in Fig. 3) in redox catalysis has been proven by electroanalytical methods. Other examples corresponding to various cases of the kinetic zone diagram have been reviewed elsewhere [18].

3.2.4 Practical Problems

Theories neglect that catalysts usually have limited turnover numbers due to destructive side reactions. This may not be so obvious in analytical experiments but it has severe consequences for large scale applications. A simple calculation can illustrate this problem: if a redox polymer with a monomer molecular weight of 400 Da and a density of 1 g cm^{-1} is considered with all redox centers addressable from the electrode and accessible to the substrate with a turnover number of 1000, then, to react 1 mmol of substrate at a 1 cm^2 electrode surface, at least 5 µmol of active catalyst centers corresponding to 2 mg of polymer, or a dry film thickness of 20 µm are required. This is 20 times more than the calculated optimum film thickness for rather favorable conditions [172].

Certainly, the same arguments apply for "chemical redox catalysis", but as discussed above, thinner films may be effective in this case. Hence, it will be reasonable to work with modified electrodes having a *large effective area* instead of thick films, i.e. three-dimensional, porous or fibrous electrodes. The notorious problem with current/potential distribution in such electrodes [184] may be overcome by the potential bias given by selective redox catalysts. Some approaches in this direction are described in the next section.

4 Preparative Scale Electrochemistry at Modified Electrodes

4.1 General Overview

The first reported electroorganic synthesis of a sizeable amount of material at a modified electrode, in 1982, was the reduction of 1,2-dihaloalkanes at *p*-nitrostyrene coated platinum electrodes to give alkenes [185]. The preparation of stilbene was conducted on a 20 µmol scale with reported turnover numbers approaching 1×10^4. The idea of mediated electrochemistry has more frequently been pursued for inorganic electrode reactions, notably the reduction of oxygen which is of eminent importance for fuel cell cathodes [186]. Almost 20 contributions on oxygen reduction at modified electrodes have been reviewed earlier [18]. Catalysts employed are mainly viologens [181, 187, 188] and quinones [189–191], giving hydrogen peroxide, as well as the metal complexes of various porphyrins [180, 192]. Especially the last species appear to fulfill the more favored 4 electron reduction to H_2O instead of the 2 electron reduction to H_2O_2. Modified electrodes prepared by electropolymerization of a tetrakis-*o*-aminophenylporphyrin were reported to proceed with at least 10^4 turnovers of redox sites [193]. Polypyrrole films can catalyze dioxygen reduction as well [100, 101]. In none of the cited works was more than analytical scale reduction documented. The potential application of modified electrodes in fuel cells was demonstrated for an acidic methanol/oxygen cell using metal tetraarylporphyrins pyrolyzed on porous carbon supports [194]. A test cell shown at the 1987 Annual Meeting of the Fachgruppe Angewandte Elektrochemie of the GDCh was working for 100 days without appreciable decrease of the cell potential.

The reduction of carbon dioxide is another of the basic electrochemical reactions that has been studied at modified electrodes. The reduction at Co or Ni phthalocyanine in acidic solution yields formic acid [195] or carbon monoxide [196]. A very high selectivity for carbon monoxide is observed at electrodes modified with an electropolymerized Rhenium vinylbipyridine complex [197].

Other reactions of small inorganic molecules are the oxidation of chloride ion at a Nafion electrode impregnated with a ruthenium oxo complex [198] and the reduction of nitrogen monoxide to ammonia at a Co phthalocyanine modified electrode [199].

Fewer examples are reported for organic electrode reactions: some alkyl halides were catalytically reduced at electrodes coated with tetrakis-*p*-aminophenylporphyrin [200], carboxylate ions are oxidized at a triarylamine polymer [201], and $Os(bipy)_3^{+}$ in a Nafion film catalytically oxidizes ascorbic acid [202]. Frequently, modified electrodes fail to give catalytic currents for catalyst substrate combinations that do work in the homogeneous case [109] even when good permeability of the film is proven [9].

As discussed before, very high turnover numbers of the catalytic site and a large active electrode area are the most important features for effective catalysis. In the following sections three relatively successful approaches are illustrated in detail, all of which make use of one or both of these parameters. A further section will deal with non-redox modified electrodes for selectivity enhancement of follow-up reactions.

4.2 Reduction of Vicinal Organic Dihalides

An irreversible reaction of the intermediate of a redox reaction will greatly facilitate redox catalysis by thermodynamic control. A good example is the reduction of the carbon halogen bond where the irreversible reaction is the cleavage of the carbon halogen bond associated, or concerted, with the first electron transfer [166, 203]. The reactive carbon-centered radicals produced in this step may however be disastrous for the polymer structure of the electrode coatings. The situation is more favorable for vicinal dihalides because of the rapid elimination of the second halide ion together with the uptake of a second electron. With specially suited redox catalysts this reductive elimination can even proceed as a concerted, "inner sphere" two-electron process [204]. Even the non-catalyzed reaction at the electrode surface has been described as "concerted" [205, 206].

It is not surprising then that the first reported surface mediated reaction made use of vicinal dihalides. In addition to poly-*p*-nitrostyrene [185], immobilized metalloporphyrins [198)207)] and viologens [208)] were found to be catalytically active. Recently, the mediated reduction of meso- and dl-dibromodiphenylethane at viologen modified electrodes was re-examined using a polymer formed directly at the electrode surface by anodic oxidation of suitable pyrrole monomers (*13*, *14*, *15*, Scheme 7) [113,210,211]. The electroanalytical proof of effective mediation was obtained by cyclic voltammetry (Fig. 4 [210)]) although no attempt of a quantitative kinetic assay was made. The catalytic reduction of dibromostilbene occurs in the second reduction peak of the viologen system, almost 1 V more positive than without mediator. It would be interesting to know if the neutral viologen species can transfer *two electrons* to the substrate.

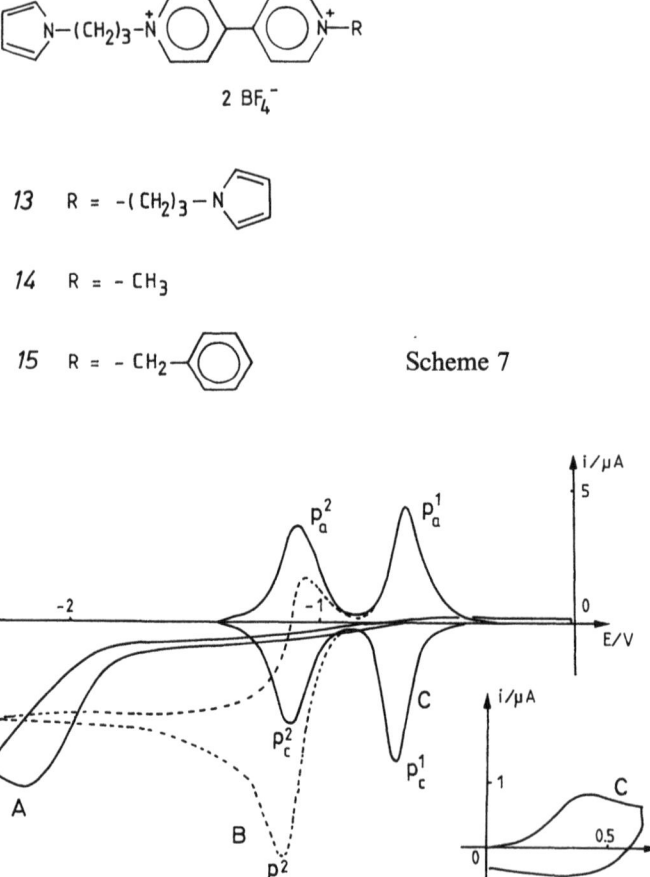

Scheme 7

Fig. 4. Catalytic reduction of *meso*-1,2-dibromo-1,2-diphenylethane (*DBDPE*) at poly-*13*-coated Pt. cyclic voltammograms in 0.1 M TBAP/acetonitrile at 0.1 V s^{-1} scan rate. A: *16* at bare Pt; B: DBDPE at poly-*13* ($\Gamma = 1 \times 10^{-8}$ mol cm^{-2}); C: poly-*13* in clean electrolyte (from ref. [210)]).

Catalytic currents for a given scan rate were compared for the three different polymers and for increasing thickness of the films. An optimum film thickness for maximum catalytic efficiency was found that was depending on the type of polymer. Poly-*14* and poly-*15* had a higher optimum thickness and catalytic efficiencies although the electrochemistry of all three polymers in pure solvent is very similar. The authors ascribed this effect to different permeabilties of the polymers. Interestingly, with very thin films (2×10^{-9} mol cm^{-2}) of poly-*13*, catalytic effects were no longer observed neither could reversible redox species in solution be detected. This was assigned to a lesser permeability of the film closer to the electrode surface than in regions stretching out further into solution. These results document that, of necessity, deviations from theory must be observed because the diffusion coefficients of substrates in polymer films may not be constant throughout the reaction layer.

Controlled potential electrolyses of dibromostilbenes at poly-*13* coated platinum foils [210,211] gave results similar to those previously observed [185]. The electrodes lose their catalytic activity after the passage of about 2 μF cm^{-2}, although up to 80% of the viologen electroactivity is retained in the film. Thus, it is obvious that redox conduction within the film is still fast and not rate determining. The absence of catalysis could be explained by either kinetic inhibition of the mediation reaction by accumulated bromide ions or, as favored by the authors, by irreversible reactions within the polymer backbone blocking substrate diffusion into the film.

Furthermore, it was found that, although the catalytic efficiency measured in initial current was lower for *13*, the latter was more stable. When *carbon felt electrodes* were coated with this polymer (approx. 1×10^{-5} mol on a $20 \times 20 \times 4$ mm piece) the reduction of dibromostilbene was obtained on a mmol scale with only moderate decrease of mediator electroactivity [211]. The production of about 0.15 mmol of stilbene is the highest hitherto reported yield at an organic polymer modified electrode.

4.3 Electrocatalytic Hydrogenation

The viologen-coated carbon felt electrode has, in fact, to be rated as a success that points the right way. Recently, in the same research group a further application of the polypyrrole based viologen electrode has been demonstrated: when electrodes as described above were additionally impregnated with palladium or rhodium micro particles, the electrochemical hydrogenation of activated alkenes, acetylenic triple bonds, aromatic aldehydes, and aromatic nitro groups was achieved in batches up to 15 mmol (Table 3 [212]). The electrolyte was aqueous potassium chloride with added ethanol or 2-ethoxyethanol to achieve solubility of the substrates. The electrochemical reaction proceeded at −0.4 to 0.5 V (vs. SCE). The electrodes appear to be very stable being reusable even after a turnover of more than 5000 molecules per Pd or Rh atom at a metal to viologen ratio of approx. 1:1.

4.4 Vitamin B$_{12}$ Reactions

Vitamin B$_{12}$ is one of the most extraordinary and effective catalysts working in biological systems [213]. The application of natural B$_{12}$ as well as its cyanocobalamine form as a redox mediator [179,214], especially in the hydrogenolysis of the carbon

Table 3. Electrocatalytic hydrogenations on carbon felt electrodes coated with polyviologen/Pd [212]

substrate	init amt, mmol	consmd current, electron molecule^{-1}	product	yield (current effency)
PhCH=CHCO$_2$H	7	2	PhCH$_2$CH$_2$CO$_2$H	95 (95)
(4-isopropylcyclohexenone)	15	2	(4-isopropylcyclohexanone)	100 (100)
(4,4-dimethyl-substituted cyclohexenone)	15	9.6	(saturated ketone)	48 (10)
PhC≡CPh	3	4	PhCH$_2$CH$_2$Ph	98 (98)
3-methoxybenzaldehyde (CH$_3$O-C$_6$H$_4$-CHO)	4	2	3-methoxybenzyl alcohol (CH$_3$O-C$_6$H$_4$-CH$_2$OH)	100 (100)
PhNO$_2$	8	6	PhNH$_2$	86 (86)

halogen bond [215] and the reductive Michael addition of alkyl and alkenyl halides to activated olefins [216] has been amply demonstrated. B$_{12}$ mediated reactions represent a typical chemical redox catalysis involving oxidative addition of alkylating agents and subsequent group transfer [179, 204, 214]. Highly reactive intermediates can be avoided in this way and high turnover numbers are likely to be achieved.

An epoxy resin made by thermal curing of an araldite prepolymer with the 3,5-diaminophenylester of a peripherically modified vitamin B$_{12}$ as "hardener" on carbon surfaces gave stable polymer films (see Scheme 3, Sect. 2.2.1). The typical B$_{12}$

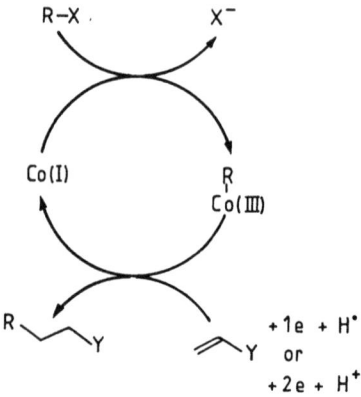

Scheme 8

electrochemistry was demonstrated by cyclic voltammetry at basal plane pyrolytic graphite electrodes [76]. The same electrode showed catalytic activity in the reductive Michael addition of ethyl iodide to acrylonitrile (Scheme 8).

Controlled potential electrolysis of the substrates, at -1.4 V vs. SCE, at a carbon felt electrode on which only 3.6×10^{-8} mol B_{12} were immobilized (approx. 1×10^{-10} mol cm^{-2}) resulted in the production of 76 µmol of valeronitrile 16, corresponding to a turnover number of 2100 [76, 217]. This example shows that the combination of inner sphere redox mediators and high surface electrodes is promising.

The reductive cleavage of the alkylcobalamine is facilitated by light irradiation and can then proceed at a much more positive potential. A demonstration photoelectrochemical reactor for the B_{12}-catalyzed photoelectrochemical synthesis of Michael adduct 17, the alarm pheromone of the ant atta texana (Scheme 9) has been constructed where the complete device is driven solely by solar energy [218]. Hopefully, mediated photoelectrochemical reactions of this type will also be realized at chemically modified electrodes.

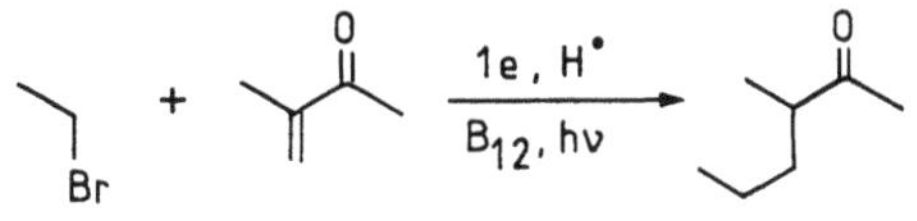

Scheme 9

4.5 Ceramic Oxide Electrodes

Many transition metal compounds are used as reduction or oxidation agents in organic chemistry. Surface modified electrodes using such reagents are well known, e.g. the alkaline nickel peroxide electrode, applications of which have been reviewed in the present electrochemistry series [219]. As an alternative approach some ceramic metal oxide electrodes (Cr_2O_3, RuO_2, MnO_2, VO_x, TiO_2) have been developed [147-149]. Typically, such electrodes combine the porosity of the oxide layers and the high reactivity of surface groups. Oxidation as well as reduction reactions can be carried out. The oxidation of 2-propanol at a Ti/Cr_2O_3 electrode is given in Scheme 10 [220].

Ceramic chromium^{III} oxide forms a green layer on the titanium surface which is oxidized to chromium trioxide in aqueous sulfuric acid. When 2-propanol is present, it is rapidly oxidized to give acetone in 100% current yield. The similarity of the molecular oxidation mechanism to the one known for the homogeneous reaction was indicated by the expected kinetic isotope effect. High turnover numbers were noted, but the operating life of the electrode is limited by leakage of Cr^{VI} into the acid solution. The possibility of stabilizing the layers by the admixture of other metal oxides is indicated [221].

Ti/TiO_2 electrodes manufactured by impregnating a Ti surface with a soluble Ti^{IV} compound and subsequent baking in air can be used for reduction processes with Ti^{III} or Ti^{II} species as proposed catalytic intermediates. The usefulness of such electrodes was demonstrated by the reduction of nitrobenzene in 1 M H_2SO_4/CH_3OH (1:1)

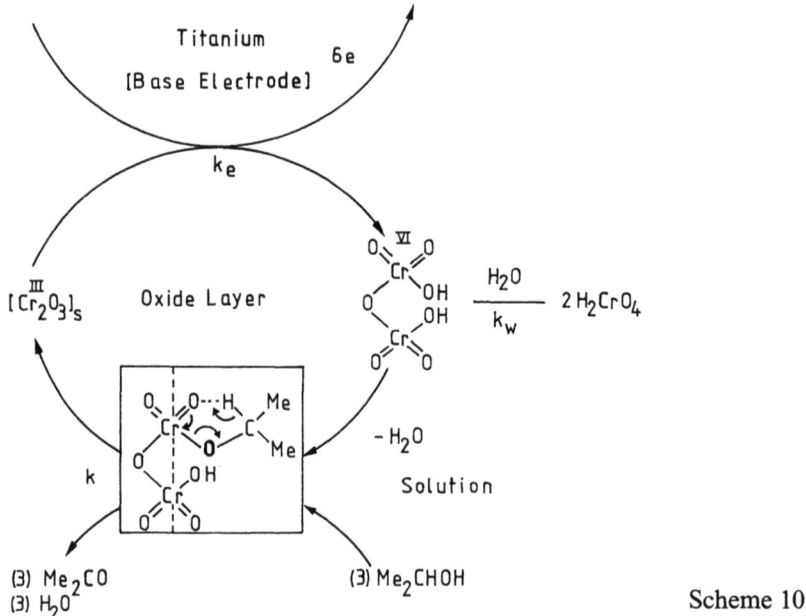

Scheme 10

with 70% current yield on a 0.4 mol scale [2222]. Since there was no hydrogen formation, other reduction products of nitrobenzene must have been formed. In both cases the oxide layers are seen as true mediator systems in a three dimensional array, and not as improved surfaces [147, 220].

4.6 Modified Electrodes for Stereoselectivity in Electroorganic Synthesis

High product selectivity is one of the most important challenges in synthetic methods. Some electrode reactions of organic substrates show a surprisingly high chemoselectivity and regioselectivity [223]. Diastereoselectivity is occasionally observed, apparently due to diastereoface discriminating adsorption at the electrode surface [224].

Enantioselective electron transfer reactions are not possible in principle because the electron cannot possess chirality. Whenever the choice of enantiodifferentiation becomes apparent, it will occur in chemical steps subsequent (or prior) to electron transfer. Thus, enantioselectivities require a chiral environment in the reaction layer of electrochemical intermediates [225] although asymmetric induction was reported for the electrochemical reduction of mandelic and pyruvic acids in the presence of a magnetic field normal to the electrode surface without any chiral inductor present [226].

Partially successful attempts towards chiral electrochemical synthesis have involved chiral supporting electrolytes [227], chiral solvents [228], and chiral adsorbates, mostly alkaloids [229–231]. With the latter method enantiometric excess values >40% have been achieved [231, 232] but these are still far from those available with modern organometallic methods and hence expected and required for a useful synthetic method [233].

From this discussion it is clear, that, independently of their redox properties, suitably modified electrodes offer themselves for the introduction of diastereo- or enantioselectivity into electrochemistry. Early reports of chiral inductions at modified electrodes include reactions at graphite [1, 234] and SnO [235] surfaces derivatized with monolayers of (S)-(—)-phenylalanine. Asymmetric inductions at the "chiral graphite electrode" [1] could, however, not be verified in other laboratories even after great efforts [236].

Subsequently, a number of reactions at poly-L-valine coated carbon electrodes [237–243] were reported to yield optically active products. Reductions, e.g. of citraconic acid [237] or 1,1-dibromo-2,2-diphenylcyclopropane [240], as well as the oxidation of aryl-alkyl sulfides [239, 242] proceeded with chiral induction at such electrodes (Table 4). Chemical and optical yields were highly dependent on the mode of fabrication of the polymer layers [241, 242]. Enantiometric excess values, chemical stability and resuability were greatly improved by using polyvaline coated on a polypyrrole film that itself was anchored to the electrode surface by a chemical derivatization method [239].

The highest reported optical yield was 93% for the oxidation of t-butyl-phenyl sulfide to the S-enantiomer of the corresponding sulfoxide [242]. The polyvaline electrode could also be used for the kinetic resolution of racemic electroactive compounds as demonstrated for the oxidation of 2,2-dimethyl-1-phenyl-1-propanol which, after work-up, left 43% pure S(—)-enantiomer following oxidation of half of the starting alcohol to the ketone [243]. Another approach of asymmetric induction at chirally modified electrodes uses the intrinsically asymmetric structure of sodium montmorillonite which has been shown to act as a chiral adsorption material for chromatography [244] and an ion exchanging electrode coating as well [136]. The potential usefulness of clay coated electrodes modified with $Ru(phen)_3^{2+}$ was shown by a kinetic resolution in the oxidation of racemic $Co(phen)_3^{2+}$ to optically active $Co(phen)_3^{3+}$ with 7% ee [245]. The asymmetric oxidation, at a similar electrode, of cyclohexyl-t-pentyl sulfide to the sulfoxide proceeded with 20% ee on a submicromolar scale [246]. The enantiomeric purity was determined from cd spectra. An interesting question is whether the asymmetric induction is imposed on the reactive complex solely by the asymmetric environment of the montmorillonite structure with the ruthenium complex acting as redox mediator, or if preferential electron transfer through one of the enantiomeric species of the complex is also important. We suggested earlier that the function of redox enzymes is representative for the ideal "chemical" redox mediator inasmuch as both electron transfer and chemical follow-up reactions are coordinated in a tailor-made microenvironment. Thus, one has to look for systems where the chirality transfer is nearly concerted with electron transfer in a real inner sphere process.

The results cited in this section indeed appear very promising and encouraging, but there are still many problems to solve. Chemical and optical yields are extremely sensitive to experimental conditions such as current density and electrolyte composition [238]. Some experimental details in the asymmetric reduction of citraconic acid are indeed puzzling, such as a temperature maximum of the optical yield, and the fact the same product enantiomer is formed regardless if D or L polyvaline was used

Table 4. Asymmetric electrochemical reactions at poly-L-valine coated carbon electrodes

Reactant	Product	Absol. config.	Yield %	% ee[a]	Ref.
ROOC-C(Me)=C(H)-COOR	(2R,3R)-dimethyl succinate diester				
R = H		R(+)	80[b]	35	237, 238)
R = CH₃		R(+)	65[b]	2.3	237, 238)
HOOC-C(Me)=C(Me)-COOH (maleic)	meso-dimethylsuccinic acid	S(−)	45[b]	10	238)
4-methylcoumarin	4-methylchroman-2-one	S(+)	8[c]	43	237)
Ph−CO−COOR	Ph−CH(OH)−COOR				
R = H		R	39[c]	0.7	240)
R = Et		S	48[c]	6.7	240)
Ph−C(=NOH)−COOH	Ph−CH(NH₂)−COOH				
R = Me		S	13[c]	6.2	240)
R = Ph		S	18[c]	2.1	240)
1,2,3-tribromo-1,2-diphenylcyclopropane	1,2-dibromo-1,2-diphenylcyclopropane	R	48[b]	17	240)
R−S−Ph	R−S(=O)−Ph				
R = Me		S	82[c]	2	242)
R = i-Pr		S	19[c]	77	242)
R = n-Bu		S	9[c]	20	242)
R = iso-Bu		S	69[c]	44	242)
R = t-Bu		S	45[c]	93	242)
R = cyclo-C₆H₆		n.d.	31[c]	51	242)
R = cyclo-C₆H₆[d]		n.d.	31[c]	22	242)

a. Optical yields are the maximum ones described within a set of various reaction conditions; current or chemical yields may be higher under different conditions; b. Current efficiency; c. Chemical yield; d. *p*-tolyl for Ph for the electrode coating [238]. Unfortunately, independent attempts [247] to reproduce asymmetric inductions in the oxidation of prochiral sulfides [239] failed to give optically active products. It thus appears that experimental procedures for asymmetric electrochemical reactions at modified electrodes are still not straightforeward enough to allow a general application [234, 247].

5 Analytical Applications

Electrochemistry has long been an important tool in analytical chemistry. Polarographic, voltammetric and coulometric techniques allow for the quantitative determination of trace levels of metal ions and electroactive organic compounds [248], amperometry is successfully used in hplc detectors [249] and the use of microelectrodes for in vitro or in vivo determination of biological materials is becoming increasingly important [22-25, 250]. A drawback of all electroanalytical methods is an intrinsic lack of substrate selectivity because many substrates, notably biologically active species, have similar redox potentials and many are difficult to discern because of Nernstian irreversibility and adorption effects at solid electrodes. With the development of ion-selective electrodes and other types of substrate selective membrane electrodes this problem has been solved very successfully [251]. An analogous approach towards substrate selective bioelectrochemistry is the use of immobilized enzyme electrodes [24, 25, 250]. Chemically modified electrodes can be designed to have similar and possibly more versatile properties and they have recently attracted wide interest in analytical chemistry [17, 22]. Polymer layers on electrodes can serve several purposes: a preconcentration medium, a substrate or class specific redox catalyst, or a substrate selective diffusion barrier. The combination of these properties can assume an unprecedented complexity approaching the action of enzyme electrodes [17].

5.1 Preconcentration Methods

Electrochemical analyses of trace level analyte concentrations can be considerably improved by forced accumulation of redox active species at, or in the vicinity of, the electrode surface from very dilute solutions. The preconcentration step can be achieved by specific binding or by a favourable partition coefficient of the analyte in electrode coatings. The subsequent voltammetric detection resembles the technique of anodic stripping [252]. All ion exchange polymers are potential matrices for the specific incorporation of ions or dipole molecules according to charge, size or hydrophilicity of the guest species. Among these polymers, sulfonated perfluorinated polymers are the most thoroughly investigated [42-48]. According to their morphology of hydrophilic and hydrophobic regions [44, 56, 57, 253], Nafion is particularly suited for the incorporation of large transition metal complex cations or organic protonated species including neurotransmitters like dopamine [22]. The specific uptake of cyclodextrines is an example for the preconcentration of polar neutral molecules [254]. The analytical principles of the method have been discussed in detail including the question of *complete* stripping for reuse of the electrodes [22, 255] and technical improvements by the use of supporting perfluorosulfonic membranes have been suggested [22]. The response times of such electrodes at rapid concentration changes have also been studied [256].

Preconcentration of metal cations is also achieved by providing ligand binding sites in polymer layers, e.g. 4-methyl-4'-vinyl-2,2'-bipyridine [257] for Fe^{2+} and Cu^{2+}. Carbon paste electrodes containing dimethylglyoxime [258, 259] or o-phenanthroline [260] have been used for trace level detection of Ni^{2+} and Cu^{2+}, respectively.

The presence of redox catalysts in the electrode coatings is not essential in the cases cited above because the entrapped redox species are of sufficient quantity to provide redox conductivity. However, the presence of an additional redox catalyst may be useful to support redox conductivity [257] or when specific chemical redox catalysis is used. An excellent example of the latter is an analytical electrode for the low level detection of alkylating agents using a vitamin B_{12} epoxy polymer on basal plane pyrolytic graphite [217]. The preconcentration step involves *irreversible* oxidative addition of R-X to the Co^I complex (see Scheme 8, Sect. 4.4). The detection by reductive voltammetry, in a two electron step, releases R^- that can be protonated in the medium. Simultaneously the original Co^I complex is restored and the electrode can be re-used. Reproducible relations between preconcentration times as well as R-X concentrations in the test solutions and voltammetric peak currents were established. The detection limit for methyl iodide is in the submicromolar range.

The electrochemical stripping of ions incorporated into polymer films can also be used in the sense of release of reagents into solution [261-264]. The electrochemically stimulated and controlled release of drugs [265, 266] and neurotransmitters [267] from doped polypyrrole films has been demonstrated. Quantitative studies were conducted using hexacyanoferrate ions [268].

The incorporation of minute amounts of certain ions into lattice type inorganic ion exchanging films can induce changes in the polymer morphology [131, 135] that can be detected electrochemically. Thus a nickel electrode modified with $[FeNi(CN)_6]^{3-}$ was used in the detection of Na^+ and Cs^+ at a 10^{-8} M level [269].

A similar approach is the use of insulating Langmuir-Blodgett films of phospholipid membranes on electrodes that can be "opened" by Ca^{2+} ions [270]. Electroactive ions can then diffuse through ion channels to the electrode surface to give a voltammetric signal. The membrane can be closed again by a sequestering agent like EDTA.

The application of modified electrodes for the assay of antibodies in serum preparations using redox indicators encapsuled into antigene marked liposomes attached to an electrode surface was suggested [17]. First model studies towards this goal [271] make use of ferricyanide ions entrapped in synthetic vesicles.

5.2 Microstructured Electrodes

Apart from molecular, or supramolecular design, modified electrodes can be improved by microstructuring. Two different strategies have been developed: multi-layered electrodes [15, 272] and the array of microelectrodes [16, 273].

5.2.1 Multilayered Electrodes

Redox or conducting polymers can be sandwiched between two electrodes either by vapor phase deposition of a porous gold film onto the outer surface of a polymer modified electrode [92, 272] or by micrometer gauge controlled contacting of the modified electrode by the counter electrode [274]. The access of counter ions from the electrolyte solution during electrochemical reactions within the redox polymer film is provided through the porous counter electrode or from the edges of the device. Sandwich electrodes of the former type were used to determine the rate of self-exchange of vicinal redox centers [272]. The second type provided an experimental method to measure the potential dependence of polypyrrole conductivity under steady state current flow [274].

The application of two successive redox polymer layers at an electrode surface gives rise to rectifying properties because the electron transport between the electrode and the outer layer has to be mediated by the inner redox polymer [91, 275]. Among several conceivable situations, the one where the inner layer possesses two reversible redox potentials (e.g. a Ru^{II}(bipy)$_3$ polymer) and the outer layer has one redox transition with a potential between the former ones (e.g. polyvinylferrocene) is most interesting [91, 275]. Such an electrode device has two opposite-sign rectifying regions beyond each threshold potential (E-Pt2 and E-Pt3) of the inner layer and an inner potential region (E-Pt2 < E-Pt1 < E-Pt3) in which the outer film can neither be oxidized nor reduced through the inner layer (Fig. 5). A bilayer lectrode sandwiched between two electrodes can be used to determine the electron transfer rate at the boundary between both redox films [276, 277].

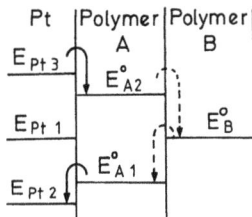

Fig. 5. Schematic representation of a Pt electrode coated with succesive layers of redox polymers A and B: a bilayer transistor electrode. Arrows indicate directions in which communication of the electrode and the outer layer is possible (from ref. [15]).

5.2.2 Microelectrode Arrays

The adaption of techniques developed for the fabrication of microelectronic components has recently led to the development of arrays of microelectrodes each of which, or groups of which can be individually addressed by an electrochemical operating instrument [273]. The microelectrodes are 6 to 8 parallel rectangular gold strips of $50 \times 2.5 \times 0.1$ µm size and 1.5 µm interelectrode gap deposited on an insulating layer of SiO$_2$ on p-silicon wafers following a photolithographic process [273, 278]. An additional layer of Si$_3$N$_4$ on top of SiO$_2$ apparently helps to avoid conductive contact between Au and Si [279]. While many new electrochemical techniques appear possible with this concept, a standard array that has been studied quite thoroughly by now is the electrochemical transistor device shown in Fig. 6 [278].

The gap between two adjacent microelectrodes (source and drain) is filled by the

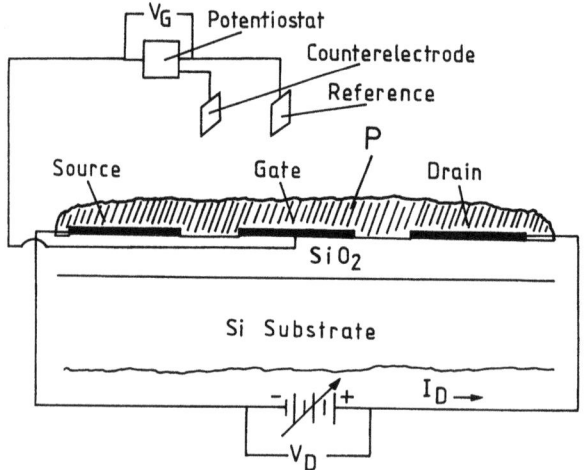

Fig. 6. Molecular transistor based on a microelectrode array. P is a polymer layer that can be switched conductive or nonconductive by the potential of the gate electrode (from ref. [273]).

deposition of an organic polymer that can be switched between a conductive and a nonconductive state. When a voltage is applied to the source-drain circuit the current flow in this circuit is controlled and possibly amplified by the redox state of the polymer the conductivity of which is manipulated by the gate circuit. Polypyrrole [273, 278], poly-3-methylthiophene [279] and polyaniline [280] have been used as connecting polymers. The transistor property of this device was shown by the induction of a 1 kHz ac signal in the source-drain circuit by a corresponding stimulation by the gate voltage [281]. The switching time as well as the amplification of the signal can be greatly reduced by decreasing the amount of bridging polymer. Thus using the technique of shadow deposition, the interelectrode gap could be reduced from 1.5 µm to 50 nm [282]. Consequently, the upper limit of the switching frequency was raised from 1 to 10 kHz. The passage of only 10^{-12} C of charge per electrode connection is necessary in each cycle.

Undoubtedly, these "devides" can still by far not compete with semiconductor electronic elements but the rapid improvement of the concept together with the potential development of better and more versatile organic polymers may allow applications in near future.

5.2.3 Sensors Based on Microelectrode Arrays

The redox processes responsible for the switching of the bridging redox polymer can also be brought about by redox processes induced by molecular species in solution [280]. Alternatively, the switching processes can be designed so that a solution component is essential for, or mediates the redox process. The array electrode can then be used as a sensor for those solution constituents.

Several demonstrations of this concept have recently been published [283-286]. The first one is based on the pH dependence of redox transitions in oxide semiconductors that are connected with conductivity changes. If the bridging polymer layer in Fig. 6 is WO_3 sputtered onto the electrode array [284], or electrochemically deposited $Ni(OH)_2$ [285], the transistor amplification is a function of the pH of the test solution. The former works at low to neutral, the latter at high pH values. In a similar way, a platinized poly(3-methylthiophene) transistor responds to pH and to dissolved hydrogen or oxygen [283]. A microelectrode array coated with poly(4-vinylpyridine) functions as a transistor when redox conductivity is activated by redox ion incorporation. Since the latter is only possible with protonated pyridine nitrogens, the function of the device is a function of both pH and the concentration of redox ion, e.g. ferricyanide [286]. The analytical applicability for all four cases was tested in flow-through cells. No interference of metal cations in the pH dependence was noted.

5.2.4 Solid State Electrochemistry

Both microstructured layer electrodes [287–289] and microelectrode arrays [290, 291] have been found to operate without a liquid electrolyte under appropriate conditions.

At a Pt-[Os(bipy)$_2$(4-vinylpyridine)$_2$](ClO$_4$)$_n$-Au sandwich electrode (see Sect. 5.2.1) plateau-like voltammograms are observed in the dry, solvent-free state when

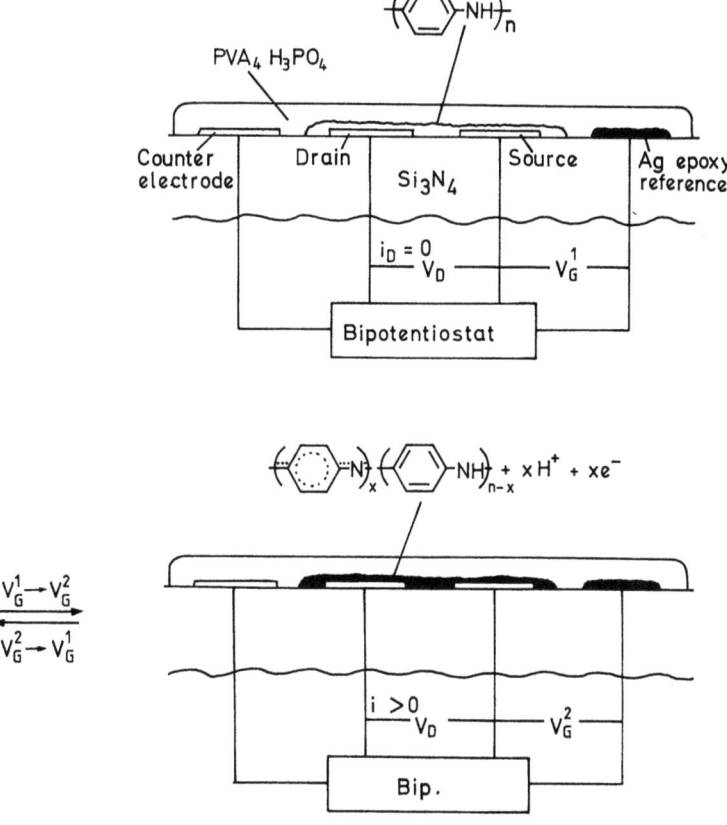

Fig. 7. Solid state molecular transistor based on polyaniline bridged microelectrodes. PVA: layer of polyvinyl alcohol; i_D: drain current; V_G^1: gate voltage rendering the polyaniline non-conductive; V_G^2: gate voltage switching on conductivity of the polyaniline layer (from ref. [291]).

a potential between the Pt and Au terminals is varied [288]. Current flow begins when the potential difference is approaching the potential gap between the $Os^{I/II}$ and $Os^{I/III}$ redox transitions. The observed conductivity is maintained by electron hopping in electron self-exchange or disproportionation processes driven by the applied potential. By electroneutrality constraints, the effect is connected to the presence of a pool of counter ions and their mobility. Thus, the polymer must have combined redox and solid state ionic conductivity. The overall process is described as electron diffusion which is an activation driven process as obtained from its temperature dependence [289].

The qualitative and quantitative appearance of the voltammograms is sensitive to the "doping level", i.e. the initial redox state of Os in the film: Os^I (n = 1), Os^{II} (n = 2) or an $Os^{II/III}$ 1:1 mixture (n = 2.5). The limiting currents are depending on the composition of the gas phase (vacuum, laboratory air, nitrogen, vapors of acetonitrile, toluene or heptane [288]). Thus, solid state electrodes of this type can possibly be used as sensors for the constituents of the surrounding atmosphere. The way these species influence the kinetics of the device is not yet clear; for instance, the effects of gases on limiting currents are partially inverted by changes in the doping level.

The microarray electrodes used for solid state electrochemistry are a slight variation of the transistor decribed in Sect. 5.2.2 [278]. The most appealing feature is the location of all the necessary electrodes on a single microchip, the reference electrode being provided by the application of a droplet of silver epoxy to one of the gold micro electrodes (Fig. 7).

The bridging polymer is a conducting poly(3-methylthiophene) [290] or polyaniline [291] and the solid state redox conduction between all electrodes is accomplished by a common coating with poly(ethyleneoxide)/$Li^+CF_3SO_3^-$ [290] or poly(vinyl alcohol)/H_3PO_4 [291]. The polyaniline based molecular transistor proved as a very sensitive moisture detector: it works well in a dry argon atmosphere but in water saturated argon the device cuts out [291].

5.3 Reference Electrodes

Reference electrodes for non-aqueous solvents are always troublesome because the necessary salt bridge may add considerable errors by undefined junction potentials. Leakage of components of the reference compartment, water in particular, into the working electrode compartment is a further problem. Whenever electrochemical cells of very small dimensions have to be designed, the construction of a suitable reference electrode system may be very difficult. Thus, an ideal reference electrode would be a simple wire introduced into the test cell. The usefulness of redox modified electrodes as reference electrodes in this respect has been studied in some detail [106, 292].

By setting the ratio of the oxidized and reduced forms of a redox couple in an electrode coating film to unity, the potential of this electrode in an inert electrolyte is poised at the half-wave potential of the couple. This has indeed been shown for platinum wires coated with polyvinylferrocene [292] or ferrocene modified polypyrrole [106]. But the long term stability of these electrodes during cell connection or under storage conditions is still insufficient for repeated use or commercial fabrication. Two reasons have to be discussed. The first is the change of the Red/Ox ratio by the Faradayic reaction at the reference electrode during measurements. With a typical surface concentration of redox centers at the modified electrode of 10^{-8} to 10^{-9} mol cm^{-2}, at very small electrodes the absolute amount may be 10^{-12} mol. With typical currents in reference circuits in the 100 pA range, a turnover of 10^{-14} mol per minute can occur that will already cause a sizeable shift of the reference potential. This argument holds for any reference couple, but such an electrode may be readjusted by potential setting.

Worse is the instability of one redox stage of the reference couple. In the ferrocene/ferrocinium system, the oxidized form is sensitive to nucleophiles and traces of water and oxygen in nonaqueous media [293] and it cannot be used in basic aqueous solutions. Typically, the potential of a ferrocene modified electrode shifts towards more positive potentials on storage, indicating a depletion of the ferrocinium form. The gross deficit of redox material is easiliy detected by voltammographic analysis of the coating. The poly(vinylferrocene) reference electrode has been recently reinvestigated with a polymer improved by crosslinking [294]. Standard potential drifts $\leq |5\text{ mV}|$ in 4 days were reported. Although the results are not satisfactory at present, research on reference systems based on polymer coated electrodes is an important task and should be continued.

6 Outlook

The field of modified electrodes spans a wide area of novel and promising research. The work cited in this article covers fundamental experimental aspects of electrochemistry such as the rate of electron transfer reactions and charge propagation within threedimensional arrays of redox centers and the distances over which electrons can be transferred in outer sphere redox reactions. Questions of polymer chemistry such as the study of permeability of membranes and the diffusion of ions and neutrals in solvent swollen polymers are accessible by new experimental techniques. There is hope of new solutions of macroscopic as well as microscopic electrochemical phenomena: the selective and kinetically facile production of substances at square meters of modified electrodes and the detection of trace levels of substances in wastes or in biological material. Technical applications of electronic devices based on molecular chemistry, even those that mimic biological systems of impulse transmission appear feasible and the construction of organic polymer batteries and color displays is close to industrial use.

Many techniques have been developed for the preparation of modified surfaces and a variety of modified electrodes have been demonstrated to function as designed. The more spectacular results have hitherto been attained for those situations where solely the electrochemistry of the coating itself is used and comparatively small currents are required. Major difficulties are encountered in most cases where redox and chemical reactions of film constituents with solute species are involved, particularly when high turnover numbers are required. All aspects of modified electrodes, however, are of high research value and require modern interdisciplinary thinking. For this reason, qualified work in this field should be greatly encouraged.

7 Note Added in Proof

A delay in the production of this volume has made it desirable to add some recent work that was published or became available to me within 1988. The arrangement of the additional material follows the headings in the main part.

2.: A newly emerging technique is the film-formation by use of self-assembling amphiphiles. Homogeneous and well adhering films are obtained by the Langmuir-Blodgett technique with films transferred from an air/water interface to an electrode surface. Examples are suitable derivatives of 2,2-bipyridine complexes [298], ferrocene [299] or 4,4'-bipyridinium salts [300]. There are also cases reported where electrode modification occurs directly by self-assembling of the amphiphiles from dilute solution to the electrode [301, 302]. The presence of a microporous structure like the alumina described in 2.3.2. [150] appears to be favorable as demonstrated for N-methyl,N'-octadecyl viologen [303]. Polymers containing amphiphilic redox groups develop structured regions depending on the redox state [304].

2.2.2.: Metal-free poly-4-vinyl-4'-methyl-2,2'-bipyridine films on electrodes have been prepared by the electroreductive polymerization of a Rh complex and subsequent leaching of the metal by a strong complexand. The films can incorporate a variety of transition metals [305].

2.2.3.: Further interesting redox modified polypyrrole films were prepared: e.g. a polymeric copper phenanthroline complex that can be reversibly de- and remetallated [306] because it retains the pseudotetrahedral environment after decomplexation. A very diversified electrochemistry is displayed by polypyrrole films containing electron donor as well as electron acceptor redox centers in the same film [307] or in adjacent layers [308]. The first electropolymerized 3-substituted pyrroles containing redox centers were also reported as copolymers with pyrrole [309, 310].

2.3.1.: New inorganic electrode films based on molybdenum and tungsten oxides have been introduced for aqueous solutions: heteropolyanion electrodes [311, 312] and $W^{V/VI}$ oxide electrodes [313]. The latter was catalytically active in the reduction of bromate ions. A polypyrrole film doped with heteropolyanions was recently reported [314]. Clay modified electrodes have been shown to be considerably stabilized by polymerization of pyrrole within the film [315, 316]. In a similar way it was found that redox molecules retained in zeolites can be activated by embedding in a carbon paste [317, 318].

2.4.: EXAFS spectroscopy was introduced as a further tool for the in situ investigation of electrode coatings [319].

3.2.: In the theoretical treatment of ion exchange polymers the roles of charge propagation [320] and of migration of ions [321, 322] were further studied by digital simulation. Another example of proven 3-dimensional redox catalysis of the oxidation of $K_3[Fe(CN)_6]$ at a ruthenium modified polyvinylpyridine coated electrode was reported [323].

4.1.: The catalytical reduction of molecular oxygen at modified electrodes remains to be the most intensely investigated topic. New examples employ several types of metal, mostly cobalt, porphyrins and phthalocyanines [302, 324–329] or viologen [317]. One of the new electrodes works in the gas phase and might be used as an oxygen sensor [324]. The reduction of carbon dioxide to methanol at various quinone modified electrodes is reported on an analytical scale [330, 331], the CO_2

reduction to CO was observed at a polypyrrole film containing a rhenium complex [332].

4.2.: The catalytic activity of viologen modified polypyrrole electrodes in preparative electroreductions [210-212] has been extended from vicinal to geminal polyhalides where the selective dehalogenation of hexachloroacetone to penta- or symm. tetrachloroacetone was achieved [333]. Preparative electrode reactions in the submillimolar region were also reported for the oxidation of benzylic alcohols at nitroxide modified polypyrrole [334] and also polyacrylic acid [335]. The epoxidation of cis-cyclooctene by molecular oxygen was electrochemically catalyzed by a manganese porphyrin attached to a polypyrrole film [336]. A polypyrrole film containing a covalently bound ruthenium bipyridine complex as well as incorporated ruthenium dioxide particles was catalytically active in the oxidation of benzylic alcohol [337].

4.6.: Some progress has also been achieved in the use of chiral polymer films at electrodes. Conductive polythiophenes containing optically active substituents in the 3-positions were prepared by electropolymerization of suitable monomers without apparent loss of optical activity [338,339]. The polymer of $R(-)$-19 exhibits distinct differences of voltammograms in the presence of either $(+)$- or $(-)$-camphor sulfonate anions [338]. Enantioselective redox transformations of the enantiomeric $Ru(bpy)_3^{2+/3+}$ complexes were detected at an electrode modified with a cholesteric liquid crystal phase [340].

R(-)-19

8 References

1. Watkins BF, Behling JR, Kariv E, Miller LL (1975) J. Am. Chem. Soc. 97: 3549
2. Grushka G (ed) (1974) Bonded Stationary Phases in Chromatography. Ann Arbor Science, Ann Arbor, Mich.
3. Moses PR, Wier L, Murray RW (1975) Anal. Chem. 47: 1882
4. Moses PR, Murray RW (1976) J. Am. Chem. Soc. 98: 7435
5. Lane RF, Hubbard AT (1973) J. Phys. Chem. 77: 1401
6. Murray RW (1980) Acc. Chem. Res. 13: 135
7. Merz A, Bard AJ (1978) J. Am. Chem. Soc. 100: 3222
8. Miller LL, Van de Mark MR (1978) J. Am. Chem. Soc. 100: 3223
9. Merz A (1982) Nachr. Chem. Techn. Lab. 30: 18
10. Murray RW (1984) in: Bard AJ (ed) Electroanalytical chemistry 13, Marcel Dekker, New York, p 191
11. Murray RW (1984) Ann. Rev. Mater. Sci. 14: 145
12. Schreurs J, Barendrecht E (1984) Recl. Trav. Chim. Pays-Bas 103: 205
13. Patriarche GJ, Vire JC, Kauffmann JM (1984) Chim. Nouv. 2: 91
14. Dong S (1986) Yingyong Huaxue 3: 1
15. Chidsey CED, Murray RW (1986) Science 231: 25
16. Wrighton MS (1986) Science 231: 33
17. Murray RW, Ewing AG, Durst RA (1987) Anal. Chem. 59: 379A

18. Hillman AR (1987) Electrochem. Sci. Technol. Polym. 1: 241
19. Degrand C (1985) Ann. Chim. (Rome) 75: 1
20. Nonaka T, Fuchigami T (1985) Yuki Gosei Kagaku Kyokaishi 43: 565
21. Fujihira M (1986) in: Fry AJ, Britton WE (eds) Topics in organic electrochemistry, Plenum, New York, p 255
22. Espenscheid MW, Ghatak-Roy AR, Moore RBIII, Penner RM, Szentirmay MN, Martin CR (1986) J. Chem. Soc., Faraday Trans. 1, 82: 1051
23. Gorton L (1986) J. Chem. Soc., Faraday Trans. 1 82: 1245
24. Lunte CE, Heinemann WR (1987) in: E. Steckhan (ed) Electrochemistry II, Top. curr. chem. 143, Springer, Berlin Heidelberg New York, p 1
25. Calabrese GS, O'Connell KM (1987) in: E. Steckhan (ed) Electrochemistry II, Top. curr. chem. 143, Springer, Berlin Heidelberg New York, p 49
26. Miller LL, VanDeMark MR (1978) J. Am. Chem. Soc. 100: 639
27. Peerce PJ, Bard AJ (1980) J. Electroanal. Chem. 112: 97
28. Gueshi T, Tohuda K, Matsuda H (1979) J. Electroanal. Chem. 101: 29
29. Andrieux CP, Haas O, Saveant JM (1986) J. Am. Chem. Soc. 108: 8175
30. Chen X, He P, Faulkner LR (1987) J. Electroanal. Chem. 222: 223
31. Itaya K, Bard AJ (1978) Anal. Chem. 50: 1487
32. Bolts JM, Wrighton MS (1979) J. Am. Chem. Soc. 101: 6179
33. Tachikawa H, Faulkner LR (1978) J. Am. Chem. Soc. 100: 4379
34. Jones ETT, Faulkner LR (1987) J. Electroanal. Chem. 222: 201
35. Oyama N, Anson FC (1980) J. Electrochem. Soc. 127: 640
36. Shigehara K, Oyama N, Anson FC (1981) J. Am. Chem. Soc. 103: 2552
37. Haas O, Vos G (1980) J. Electroanal. Chem. 113: 139
38. Haas O, Kosens M, Vos G (1981) J. Am. Chem. Soc. 103: 1331
39. Haas O, Müller N, Gerischer H (1982) Electrochim. Acta 27: 991
40. Haas O, Zumbrunnen HL, Vos JG (1985) Electrochim. Acta 30: 1551
41. Jasinsky R (1983) J. Electrochem. Soc. 130: 834
42. Rubinstein I, Bard AJ (1980) J. Am. Chem. Soc. 102: 6641
43. Rubinstein I, Bard AJ (1981) J. Am. Chem. Soc. 103: 5007
44. Henning TP, White HS, Bard AJ (1981) J. Am. Chem. Soc. 103: 3937
45. Buttry DA, Anson FC (1982) J. Am. Chem. Soc. 104: 4828
46. Buttry DA, Anson FC (1983) J. Am. Chem. Soc. 105: 685
47. Buttry DA, Saveant JM, Anson FC (1984) J. Phys. Chem. 88: 3086
48. Elliot CM, Redepenning JG (1984) J. Electroanal. Chem. 181: 137
49. Faulkner LR, Majda M (1982) J. Electroanal. Chem. 137: 149
50. Faulkner LR, Majda M (1984) J. Electroanal. Chem. 169: 77
51. Faulkner LR, Majda M (1984) J. Electroanal. Chem. 169: 97
52. Anson FC, Ohsaka T, Saveant JM (1983) J. Phys. Chem. 87: 640
53. Anson FC, Saveant JM, Shigehara K (1983) J. Am. Chem. Soc. 105: 1096
54. Oyama K, Anson FC (1980) J. Electrochem. Soc. 127: 247
55. Oyama N, Shimomura T, Shigehara K, Anson FC (1980) J. Electroanal. Chem. 112: 271
56. Rubinstein I (1984) J. Electroanal. Chem. 176: 359
57. Rubinstein I (1985) J. Electroanal. Chem. 188: 227
58. Montgomery DD, Tsuchida E, Shigehara K, Anson FC (1984) J. Am. Chem. Soc. 106: 7991
59. Montgomery DD, Anson FC (1985) J. Am. Chem. Soc. 107: 3431
60. Lindholm B, Sharp M (1986) J. Electroanal. Chem. 198: 37
61. Akahoshi H, Toshima S, Itaya K (1981) J. Phys. Chem. 85: 818
62. Oyama N, Sato K, Matsuda H (1985) J. Electrochem. Soc. 132: 1871
63. Doblhofer K, Nölte O, Ulstrup J (1978) Ber. Bunsenges. Phys. Chem. 82: 403
64. Daum P, Lenhard JR, Rolison DR, Murray RW (1980) J. Am. Chem. Soc. 102: 4649
65. Nowak RJ, Schulz PA, Umane M, Lam R, Murray RW (1980) Anal. Chem. 52: 315
66. Bowden EF, Dautartas MF, Evans JF (1987) J. Electroanal. Chem. 219: 49
67. Bowden EF, Dautartas MF, Evans JF (1987) J. Electroanal. Chem. 219: 71
68. Bowden EF, Dautartas MF, Evans JF (1987) J. Electroanal. Chem. 219: 91
69. Doblhofer K, Dürr W, Jauch M (1982) Electrochim. Acta 27: 677

70. Wöhrle D (1983) Adv. Polym. Sci. 50: 45
71. Wöhrle D, Bannehr R, Schumann B, Meyer G, Jäger N (1983) J. Mol. Catal. 21: 255
72. Wöhrle D, Bannehr D, Schumann B, Jäger N (1983) Angew. Makromol. Chem. 117: 103
73. Schumann B, Wöhrle D, Jäger N (1985) J. Electrochem. Soc. 132: 2144
74. Wöhrle D, Schumann B, Schmidt V, Jäger N (1987) Makromol. Chem., Macromol. Symp. 8: 195
75. Anson FC, Ohsaka T, Saveant JM (1983) J. Am. Chem. Soc. 105: 4883
76. Ruhe A, Walder L, Scheffold R (1985) Helv. Chim. Acta 68: 1301
77. Dominey RN, Lewis NS, Bruce JA, Bookbinder DC, Wrighton MS (1982) J. Am. Chem. Soc. 104: 4671
78. Bruce T, Murahashi M, Wrighton MS (1982) J. Phys. Chem. 86: 1152
79. Simon RA, Mallouk TE, Daube KA, Wrighton MS (1985) Inorg. Chem. 24: 3119
80. Dineen E, Schwan TC, Wilson CL (1949) J. Electrochem. Soc. 96: 22
81. Olaj OF (1987) Makromol. Chem., Macromol. Symp. 8: 235
82. Otero TF, Ponce MT, Jimenez YT (1987) Makromol. Chem., Macromol. Symp. 8: 255
83. Mengoli G, Musiani MM (1987) Makromol. Chem., Macromol. Symp. 8: 265
84. Beck F, Gunder H (1987) Makromol. Chem., Macromol. Symp. 8: 285
85. Pham MC, Lacaze PC, Dubois JE (1978) J. Electroanal. Chem. 86: 147
86. Volkov A, Tourillon G, Lacaze PC, Dubois JE (1980) J. Electroanal. Chem. 115: 279
87. Dubois JE, Lacaze PC, Pham MC (1981) J. Electroanal. Chem. 117: 233
88. Pham MC, Dubois JE, Lacaze PC (1983) J. Electroanal. Chem. 130: 346
89. Pham MC, Lacaze PC, Dubois JE (1986) Bull. Soc. Chim. Fr. p 162
90. Ghosh PK, Spiro TG (1981) J. Electrochem. Soc. 127: 654
91. Denisewich P, Willman KW, Murray RW (1981) J. Am. Chem. Soc. 103: 4727
92. Pickup PG, Kutner W, Leidner CR, Murray RW (1984) J. Am. Chem. Soc. 106: 1991
93. Potts KT, Usifer DA, Guadalupe A, Abruna HD (1987) J. Am. Chem. Soc. 109: 3961
94. Diaz AF, Kanazawa KK, Gardini GP (1979) J. Chem. Soc., Chem. Comm. p 635
95. Skotheim TA (ed) (1986) Handbook of conducting polymers, vol I and II. Marcel Dekker, New York
96. Heinze J (1988) this volume
97. Saraceno RA, Pack JG, Ewing AG (1986) J. Electroanal. Chem. 197: 265
98. Jakobs RCM, Janssen LJJ, Barendrecht E (1985) Electrochim. Acta 30: 1313
99. Haimerl A, Merz A (1987) J. Electroanal. Chem. 220: 55
100. Ikeda O, Okabayashi K, Yoshida N, Tamura H (1985) J. Electroanal. Chem. 191: 157
101. Jakobs RCM, Janssen LJJ, Barendrecht E (1985) Electrochim. Acta 30: 1433
102. Saloma, M Aguilar M, Salmon M (1985) J. Electrochem. Soc. 132: 2379
103. Rosenthal MV, Skotheim TA, Linkous C, Florit MI (1984) Polym. Prep. (ACS Div. Polym. Chem.) 25: 258
104. Rosenthal MV, Skotheim TA, Melo A, Florit MI (1985) J. Electroanal. Chem. 185: 297
105. Reynolds JR, Poropatic PH, Toyooka RL (1987) Macromolecules 20: 958
106. Haimerl A, Merz A (1986) Angew. Chem. 98: 179; Angew. Chem. Int. Ed. Engl. 25: 180
107. Merz A, Baumann R, Haimerl A (1987) Makromol. Chem., Macrom. Symp. 8: 61
108. Bidan G, Limosin D (1986) Ann. Phys. (Paris) 11 (1. Suppl. Journ. Films Org. Modif. Surf. Propr. Induites), p 5
109. Merz A, Baumann R, Haimerl A (1988) DECHEMA Monogr., in press
110. Cosnier S, Deronzier A, Moutet JC (1986) Ann. Phys. (Paris) 11 (1. Suppl. Journ. Films Org. Modif. Surf. Propr. Induites) p 113
111. Bidan G, Deronzier A, Moutet JC (1984) Nouv. J. Chim. 8: 501
112. Eaves JG, Munro HS, Parker D (1987) Inorg. Chem. 26: 644
113. Bidan G, Deronzier A, Moutet JC (1984) J. Chem. Soc., Chem. Comm. p 1185
114. Audebert P, Bidan G, Lapkowski M (1987) J. Electroanal. Chem. 219: 165
115. Bettelheim A, White BH, Raybeck SA, Murray RW (1987) Inorg. Chem. 26: 1009
116. Deronzier A, Latour JM (1987) J. Electroanal. Chem. 224: 295
117. Diaz AF, Castillo JI, Logan JA, Lee WY (1981) J. Electroanal. Chem. 129: 115
118. Diaz AF, Martinez A, Kanazawa KK, Salmon M (1981) J. Electroanal. Chem. 130: 181
119. Merz A, Haimerl A, Owen AJ (1988) Synth. Met. 25: 89
120. Bull FA, Fan FR, Bard AJ (1983) J. Electrochem. Soc. 130: 1636

121. Skotheim TA, Rosenthal MV, Linkous CA (1985) J. Chem. Soc., Chem. Comm. p 612
122. Umana M, Waller J (1986) Anal. Chem. 58: 2979
123. Bartlett RC, Whitaker RG (1987) J. Electroanal. Chem. 224: 37
124. Fan FR, Bard AJ (1986) J. Electrochem. Soc. 133: 301
125. Penner RM, Martin CR (1986) J. Electrochem. Soc. 133: 310
126. Shimidzu T, Ohtani A, Iyoda T, Honda K (1986) J. Chem. Soc., Chem. Common p 1415
127. Iyoda T, Ohtani A, Shimidzu T, Honda K (1987) Synth. Met. 18: 757
128. Yoneyama H, Takayuki H, Kuwabata B, Ikeda O (1986) Chem. Lett. p 1243
129. Mizutani F, Iijima S, Tanabe Y, Tsuda K (1985) J. Chem. Soc., Chem. Comm. p 1728
130. Mizutani F, Iijima S, Tanabe Y, Tsuda K (1987) Synth. Met. 18: 111
131. Itaya K, Uchida I, Neff VD (1986) Acc. Chem. Res. 19: 162
132. Neff VD (1978) J. Electrochem. Soc. 125: 866
133. Itaya K, Akahoshi H, Toshima S (1982) J. Electrochem. Soc. 129: 1498
134. Itaya K, Akahoshi H, Toshima S (1982) J. Appl. Phys. 53: 804
135. Itaya K, Ataka T, Toshima S, Shinohara T (1982) J. Phys. Chem. 86: 2415
136. Ghosh PK, Bard AJ (1983) J. Am. Chem. Soc. 105: 5691
137. Ghosh PK, Man AW, Bard AJ (1984) J. Electroanal. Chem. 169: 315
138. Ege D, Ghosh PK, White JR, Eqey JF, Bard AJ (1985) J. Am. Chem. Soc. 107: 5644
139. White JR, Bard AJ (1986) J. Electroanal. Chem. 197: 233
140. Itaya K, Bard AJ (1985) J. Phys. Chem. 89: 5565
141. Rudzinsky WE, Bard AJ (1986) J. Electroanal. Chem. 199: 323
142. Lin HY, Anson FC (1985) J. Electroanal. Chem. 184: 411
143. Yamagishi A, Aramata I (1984) J. Chem. Soc., Chem. Comm. p 452
144. Cornelis A, Laszlo P (1985) Synthesis, p 909
145. Murray CG, Nowak RJ, Rolison DR (1984) J. Electroanal. Chem. 164: 205
146. Li Z, Mallouk TE (1987) J. Phys. Chem. 91: 643
147. Beck F, Schulz H (1984) Ber. Bunsenges. Phys. Chem. 88: 155
148. Beck F, Gabriel W (1985) J. Electroanal. Chem. 182: 355
149. Beck F, Wermeckes B, Zimmer E (1988) DECHEMA-Monogr. (in press)
150. Miller CJ, Majda M (1985) J. Am. Chem. Soc. 107: 1419
151. Lundgren CA, Murray RW (1987) J. Electroanal. Chem. 227: 287
152. Woodruff DP, Delcher TA (1986) Modern techniques of surface science. Cambridge University Press, New York
153. Buschmann HW, Wilhelm S, Vielstich W (1986) Electrochim. Acta 31: 939
154. Fleischmann M, Oliver A, Robinson J (1986) Electrochim. Acta 31: 899
155. Farquharson S, Milner D, Tadoyyoni MA, Weaver MJ (1984) J. Electroanal. Chem. 178: 143
156. Lui NY, Fan FRF, Liu CW, Bard AJ (1986) J. Am. Chem. Soc. 108: 3838
157. Steckhan E in: Steckhan E (ed) Electrochemistry I, Top. curr. chem. 142, Springer, Berlin Heidelberg New York, p 1
158. Andrieux CP, Dumas-Bouchiat JM, Saveant JM (1978) J. Elelectroanal. Chem. 87: 39
159. Wienkamp R, Steckhan E (1982) Angew. Chem. 94: 786; Angew. Chem. Suppl. p. 1739; Angew. Chem. Int. Ed. Engl. 21: 782
160. Ruppert R, Herrmann S, Steckhan E (1987) Tetrahedron Lett. 28: 6583; Franke M, Steckhan E (1988) Angew. Chem. 100: 295; Angew. Chem. Int. Ed. Engl. 27: 265
161. Bergbreiter DE, Chandran R (1987) J. Am. Chem. Soc. 109: 174
162. Andrieux CP, Dumas-Bouchiat JM, Saveant JM (1980) J. Electroanal. Chem. 113: 9
163. Andrieux CP, Blocman C, Dumas-Bouchiat JM, M'Halla F, Saveant JM (1980) J. Am. Chem. Soc. 102: 3806
164. Evans DH, Xie N (1983) J. Am. Chem. Soc. 105: 315
165. Andrieux CP, Merz A, Saveant JM, Tomahogh R (1984) J. Am. Chem. Soc. 106: 1957
166. Andrieux CP, Merz A, Saveant JM (1985) J. Am. Chem. Soc. 107: 6097
167. Andrieux CP, Saveant JM (1978) J. Electroanal. Chem. 93: 163
168. Andrieux CP, Dumas-Bouchiat JM, Saveant JM (1982) J. Electroanal. Chem. 131: 1
169. Andrieux CP, Saveant JM (1982) J. Electroanal. Chem. 142 :1
170. Andrieux CP, Saveant JM (1984) J. Electroanal. Chem. 169: 9
171. Albery WJ, Hillman RA (1981) Roy. Soc. Chem. Ann. Rep. 78: 377

172. Albery WJ, Hillman RA (1984) J. Electroanal. Chem. 170: 27
173. Leddy J, Bard AJ, Maloy T, Saveant JM (1985) J. Electroanal. Chem. 187: 205
174. Andrieux CP (1986) in: Smyth MR, Vos JG (eds) Electrochemistry, sensors and analysis, Anal. Chem. Symp. Ser. 25, Elsevier, Amsterdam, p 225
175. Anson FC (1980) J. Phys. Chem. 84: 3336
176. Kaufman FB, Schroeder AH, Engler EM, Kramer SR, Chambers JQ (1980) J. Am. Chem. Soc. 102: 1483
177. Facci RH, Schmehl RH, Murray RW (1982) J. Am. Chem. Soc. 104: 4959
178. Nakahama S, Murray RW (1983) J. Electroanal. Chem. 158: 303
179. Lexa D, Saveant JM, Soufflet J (1979) J. Electroanal. Chem. 100: 159
180. Anson FC, Ni CL, Saveant JM (1985) J. Am. Chem. Soc. 107: 3442
181. Oyama N, Anson FC (1980) Anal. Chem. 52: 1192
182. Anson FC, Saveant JM, Shigehara K (1983) J. Electroanal. Chem. 145: 423
183. Oyama N, Oki N, Ohno H, Ohnuki Y, Matsuda H, Tsuchida E (1983) J. Phys. Chem. 87: 3642
184. Kreysa G (1983) in: Behrens D, Kreysa G (eds) DECHEMA-Monogr. 94, Verlag Chemie, Weinheim, p 123
185. Kerr JB, Miller LL (1979) J. Electroanal. Chem. 101: 263
186. Kordesch K (1984) Brennstoffbatterien, Springer, Vienna New York
187. Martigny P, Anson FC (1982) J. Electroanal. Chem. 139: 383
188. Janda P, Weber J, Kavan L (1984) J. Electroanal. Chem. 180: 109
189. Calabrese GS, Buchanan RM, Wrighton MS (1983) J. Am. Chem. Soc. 105: 5594
190. Degrand C (1984) J. Electroanal. Chem. 169: 259
191. Holdcroft S, Funt LB (1987) J. Electroanal. Chem. 255: 177
192. Buttry DA, Anson FC (1984) J. Am. Chem. Soc. 106: 59
193. Bettelheim A, White BA, Murray RW (1987) J. Electroanal. Chem. 217: 271
194. Ahn J, Holze R, Vielstich W (1988) DECHEMA-Monogr. (in press)
195. Kapusta S, Hackerman N (1984) J. Electrochem. Soc. 131: 1511
196. Lieber C, Lewis NS (1984) J. Am. Chem. Soc. 106: 5033
197. Toole TR, Margerum LD, Westmoreland TD, Vining WJ, Murray RW (1985) J. Chem. Soc., Chem. Comm. p 1416
198. Vining J, Meyer TI (1985) J. Electroanal. Chem. 195: 183
199. Ogura K, Yamasaki S (1981) J. Appl. Electrochem. 15: 279
200. Elliot CM, Marrese CA (1981) J. Electroanal. Chem. 119: 395
201. Moutet JC (1984) J. Electroanal. Chem. 161: 181
202. Anson FC, Tsou YM, Saveant JM (1984) J. Electroanal. Chem. 178: 13
203. Saveant JM (1987) J. Am. Chem. Soc. 109: 6788
204. Lexa D, Seavant JM, Wang DL, Su KB (1987) J. Am. Chem. Soc. 109: 6464
205. Casanova J, Rogers·HR (1974) J. Org. Chem. 39: 2408
206. Merz A (1977) Electrochim. Acta 22: 1271
207. Kerr JB, Miller LL, Van de Mark MR (1980) J. Am. Chem. Soc. 102: 3383
208. Rocklin RD, Murray RW (1981) J. Phys. Chem. 85: 2104
209. Willman KW, Murray RW (1982) J. Electroanal. Chem. 133: 211
210. Coche L, Deronzier A, Moutet JC (1986) J. Electroanal. Chem. 198: 187
211. Coche L, Moutet JC (1987) J. Electroanal. Chem. 224: 111
212. Coche L, Moutet JC (1987) J. Am. Chem. Soc. 109: 6887
213. Schneider Z, Stroinsky A (1987) Comprehensive B_{12}. W. de Gruyter, Berlin
214. Scheffold R, Rytz G, Walder L (1983) Scheffold R (ed) Modern synthetic methods vol 3, Salle, Frankfurt and Verlag Sauerländer, Aarau, p 355
215. Walder L, Rytz G, Meier K, Scheffold R (1978) Helv. Chim. Acta 61: 3013
216. Scheffold R (1985) Chimia 39: 203
217. Ruhe A, Walder L, Scheffold R (1987) Makromol. Chem., Macromol. Symp. 8: 225
218. Steiger B, Walder L, Scheffold R (1986) Chimia 40: 93
219. Schäfer H (1987) in: Steckhan E (ed) Electrochemistry I, Top. curr. chem. 142, Springer, Berlin Heidelberg New York, p 101
220. Schulz H, Beck F (1985) Angew. Chem. 97: 1047; Angew. Chem. Int. Ed. Engl. 24: 1049

221. Beck F, Schulz H (1984) Electrochim. Acta 29: 1569
222. Beck F, Gabriel W (1985) Angew. Chem. 97: 765; Angew. Chem. Int. Ed. Engl. 24: 771
223. Beck F (1974) Electroorganische Chemie, Verlag Chemie, Weinheim
224. Anderson JT, Stocker JH (1983) in: Baizer MM, Lund H (eds) Organic electrochemistry, Marcel Dekker, New York, p 905
225. Tallec A (1985) Bull. Soc. Chim. Fr. p 743
226. Takahashi F, Tomii K, Takahashi H (1986) Electrochim. Acta 31: 127
227. Horner L (1983) in: Baizer MM, Lund H (eds) Organic electrochemistry, Marcel Dekker, New York, p 945
228. Seebach D, An Oei H (1975) Angew. Chem. 87: 629; Angew. Chem. Int. Ed. Engl. 14: 634
229. Gourley RN, Grimshaw J, Millar PG (1970) J. Chem. Soc. (C) p 2318
230. Jubault M, Raoult E, Peltier D (1974) Electrochim. Acta 19: 865
231. Kopilov J, Kariv E, Miller LL (1977) J. Am. Chem. Soc. 99: 3450
232. Hazard R, Jaouannet S, Tallec A (1982) Tetrahedron 38: 92
233. Nogradi M (1986) Stereoselective synthesis. VCH Verlagsges., Weinheim
234. Firth BE, Miller LL, Milani M, Rogers T Lennox J, Murray RW (1976) J. Am. Chem. Soc. 98: 8271
235. Firth BE, Miller LL (1976) J. Am. Chem. Soc. 98: 8272
236. Horner L, Brich W (1977) Liebigs Ann. Chem. p 1354
237. Abe S, Nonaka T, Fuchigami T (1983) J. Am. Chem. Soc. 105: 3630
238. Nonaka T, Abe S, Fuchigami T (1983) Bull. Soc. Chem. Jpn. 56: 2778
239. Komori T, Nonaka T (1983) J. Am. Chem. Soc. 105: 5690
240. Abe S, Fuchigami T, Nonaka T (1983) Chem. Lett. p 1033
241. Abe S, Nonaka T (1983) Chem. Lett. p 1541
242. Komori T, Nonaka T (1984) J. Am. Chem. Soc. 106: 2656
243. Komori T, Nonaka T (1984) Chem. Lett. p 509
244. Yamagishi A (1985) J. Am. Chem. Soc. 107: 732
245. Yamagishi A, Aramata A (1984) J. Chem. Soc., Chem. Comm. p 452
246. Yamagishi A, Aramata A (1985) J. Electroanal. Chem. 191: 449
247. Decker M, Schäfer HJ (1987) in: Lund H (ed) Abstract of Papers, XIII. Sandbjerg Meeting on Organic Electrochemistry, University of Aarhus, p 29
248. Plambeck JA (1982) Electroanalytical chemistry, John Wiley, New York
249. Weber SG (1986) in: Yeung ES (ed) Detectors for liquid chromatography, John Wiley, New York, p 229
250. Frew JE, Hill HAO (1987) Anal. Chem. 59: 933A
251. Bailey PL (1980) Analysis with ion-selective electrodes, Heyden, London
252. Barendrecht E (1967) in: Bard AJ (ed) Electroanal. Chem. 2, Marcel Dekker, New York, p 53
253. Vining WJ, Meyer TJ (1987) J. Electroanal. Chem. 237: 191
254. Matsue T, Akiba O, Osa T (1986) Anal. Chem. 58: 2096
255. Whiteley LD, Martin CR (1987) Anal. Chem. 59: 1746
256. Kristensen EW, Kuhr WG, Wightman RM (1987) Anal. Chem. 59: 1757
257. Guadalupe AR, Abruna HD (1985) Anal. Chem. 57: 142
258. Baldwin RP, Christensen DK, Kryger L (1986) Anal. Chem. 58: 1790
259. Thomson KN, Kryger L, Baldwyn RP (1988) Anal. Chem. 60: 151
260. Prabhu, SV, Baldwin RP, Kryger L (1987) Anal. Chem. 59: 1054
261. Miller LL, Lau AKN, Miller E (1982) J. Am. Chem. Soc. 104: 5242
262. Lau AKN, Miller LL (1983) J. Am. Chem. Soc. 105: 5271
263. Lau AKN, Miller LL, Zinger B (1983) J. Am. Chem. Soc. 105: 5278
264. Lau AKN, Zinger B, Miller LL (1983) Neurosci. Lett. 35: 101
265. Zinger B, Miller LL (1984) J. Am. Chem. Soc. 106: 6861
266. Blankespoor R, Miller LL (1985) J. Chem. Soc., Chem. Comm., p 90
267. Zinger B, Miller LL (1984) J. Electroanal. Chem. 181: 153
268. Miller LL, Zinger B, Zhou QX (1987) J. Am. Chem. Soc. 109: 2267

269. Amos LJ, Duggal A, Mirsky EJ, Ragonesi P, Bocarsly AB, Fitzgerald-Bocarsly PA (1988) Anal. Chem. 60: 245
270. Sugawara M, Kojima K, Sazawa H, Umezawa Y (1987) Anal. Chem. 59: 2842
271. Kannuck RM, Bellama JM, Durst RA (1988) Anal. Chem. 60: 142
272. Pickup PG, Murray RW (1983) J. Am. Chem. Soc. 105: 4510
273. White HS, Kittlesen GP, Wrighton MS (1984) J. Am. Chem. Soc. 106: 5375
274. Feldman BJ, Burgmayer P, Murray RW (1985) J. Am. Chem. Soc. 107: 872
275. Abruna HD, Denisevich P, Murray RW (1981) J. Am. Chem. Soc. 103: 1
276. Pickup PG, Murray RW (1984) J. Electrochem. Soc. 131: 833
277. Leidner CR, Murray RW (1985) J. Am. Chem. Soc. 107: 551
278. Kittlesen GP, White HS, Wrighton MS (1984) J. Am. Chem. Soc. 106: 7389
279. Thackeray JW, White HS, Wrighton MS (1985) J. Phys. Chem. 89: 5133
280. Paul EW, Ricco AJ, Wrighton MS (1985) J. Phys. Chem. 89: 1441
281. Lofton EP, Thackeray WT, Wrighton MS (1986) J. Phys. Chem. 90: 6080
282. Jones ETT, Chyan OM, Wrighton MS (1987) J. Am. Chem. Soc. 109: 5526
283. Thackeray JW, Wrighton MS (1986) J. Phys. Chem. 90: 6674
284. Natan MJ, Mallouk TE, Wrighton MS (1987) J. Phys. Chem. 91: 648
285. Natan JN, Belanger D, Carpenter MK, Wrighton MS (1987) J. Phys. Chem. 91: 1834
286. Belanger D, Wrighton MS (1987) Anal. Chem. 59: 1426
287. Jernigan JC, Chidsey CED, Murray RW (1985) J. Am. Chem. Soc. 107: 2824
288. Jernigan JC, Murray RW (1987) J. Am. Chem. Soc. 109: 1738
289. Jernigan JC, Murray RW (1987) J. Phys. Chem. 91: 2030
290. Chao S, Wrighton MS (1987) J. Am. Chem. Soc. 109: 2197
291. Chao S, Wrighton MS (1987) J. Am. Chem. Soc. 109: 6627
292. Peerce PJ, Bard AJ (1980) J. Electroanal. Chem. 108: 221
293. Sato M, Yamada T, Nishikama A (1980) Chem. Lett. p 925
294. Kannuck RM, Bellama JM, Blubaugh EA, Durst RA (1987) Anal. Chem. 59: 1473
295. Bedioui F, Bongars C, Devynck J, Bied-Charrenton C, Hinnen C (1986) J. Electroanal. Chem. 207: 87
296. Daire F, Bedioui F, Devynck J, Bied-Charrenton C (1987) J. Electroanal. Chem. 224: 95
297. Bedioui F, Merino A, Devynck J, Mestres CE, Bied-Charrenton C (1988) J. Electroanal. Chem. 239: 89
298. Daifuhu H, Yoshimura K, Aoki K, Tokuda K, Matsuda H (1986) J. Electroanal. Chem. 199: 47
399. Facci JS, Falcino PA, Gold JM (1986) Langmuir 2: 732
300. Lee CW, Bard AJ (1988) J. Electroanal. Chem. 239: 441
301. Diaz A, Kaifer AE (1988) J. Electroanal. Chem. 249: 333
302. Van Galen DA, Majda M (1988) Anal. Chem. 60: 1549
303. Miller CJ, Majda M (1988) Anal. Chem. 60: 1168
304. Tsou YM, Liu HY, Bard AJ (1988) J. Electrochem. Soc. 135: 1669
305. Meyer TJ, Sullivan BP, Caspar JV (1987) Inorg. Chem. 25: 4145
306. Bidan G, Divisia-Blohorn B, Kern J-M, Sauvage J-P (1988) J. Chem. Soc., Chem. Comm. p 723
307. Deronzier A, Essakalli M, Moutet JC (1988) J. Electroanal. Chem. 244: 163
308. Downward AJ, Surridge NA, Meyer TJ, Cosnier S, Deronzier A, Moutet JC (1988) J. Electroanal. Chem. 246: 321
309. Inagaki T, Hunter M, Yang XQ, Skotheim TA, Okamoto Y (1988) J. Chem. Soc., Chem. Comm. p 126
310. Inagaki T, Hunter M, Yang XQ, Skotheim TA, Lee HS, Okamoto Y (1988) Mol. Cryst. Liq. Cryst. 160: 79
311. Keita B, Nadjo L (1987) J. Electroanal. Chem. 230: 267
312. Keita B, Nadjo L, Saveant JM (1988) J. Electroanal. Chem. 243: 267
313. Kulesza PJ, Faulkner LR (1988) J. Am. Chem. Soc. 110: 4905
314. Kulesza PJ, Faulkner LR (1988) J. Electroanal. Chem. 248: 305
315. Castro-Acuna CM, Fan FR, Bard AJ (1987) J. Electroanal. Chem. 234: 347
316. Rudzinski WE, Figueroa C, Hoppe C, Kuramoto TY, Root D (1988) J. Electroanal. Chem. 243: 367

317. Creasy KE, Shaw BR (1988) Electrochim. Acta 33: 551
318. Shaw BR, Creasy KE, Lanczycki CJ, Sargeant JA, Tirhado M (1988) J. Electrochem. Soc. 135: 869
319. Albarelli MJ, White JH, Bommarito GM, McMillan M, Abruna HD (1988) J. Electroanal. Chem. 248: 77
320. Lindhom B (1988) J. Electroanal. Chem. 251: 297
321. Lange R, Doblhofer K (1987) J. Electroanal. Chem. 237: 13
322. Lange R, Doblhofer K (1988) Ber. Bunsen-Ges. Phys. Chem. 92: 578
323. Cassidy JF, Vos JG (1988) J. Electrochem. Soc. 135: 863
324. Bettelheim A, Reed R, Hendricks NH, Collman JP, Murray RW (1987) J. Electroanal. Chem. 238: 259
325. O'Brien RN, Santhanam KSV (1987) Electrochim. Acta 32: 1209
326. Paliteiro C, Hamnett A, Goodenough JB (1988) J. Electroanal. Chem. 239: 273; ibid. 249: 167
327. Jiang IL, Dong S (1988) J. Electroanal. Chem. 246: 101
328. Dong S, Kuwana T (1988) Electrochim. Acta 33: 667
329. Collman JP, Hendricks NH, Leidner CR, Ngameni F, L'Her M (1988) Inorg. Chem. 27: 387
330. Ogura K, Fujita M (1987) J. Mol. Catal. 41: 303
331. Ogura K, Yoshida I (1987) Electrochim. Acta 32: 1191
332. Cosnier S, Deronzier, Moutet JC (1988) J. Mol. Catal. 45: 381
333. Coche L, Moutet JC (1988) J. Electroanal. Chem. 245: 313
334. Deronzier A, Limosin D, Moutet JC (1987) Electrochim. Acta 32: 1643
335. Osa T, Akiba U, Segawa I (1988) Chem. Lett. p 1423
336. Moisy P, Bedioui F, Robin Y, Devynck J (1988) J. Electroanal. Chem. 250: 191
337. Cosnier S, Deronzier A, Moutet JC (1988) Inorg. Chem. 27: 2389
338. Lemaire M, Delabouglise D, Garreau R, Guy R, Roncali J (1988) J. Chem. Soc., Chem. Comm. p. 658
339. Dilip K, Vishwas J, Pushpito KG (1988) J. Chem. Soc., Chem. Comm. p 917
340. Mariani RD, Abruna HD (1987) Gov. Rep. Announce. Index (U.S.) 87(23): Abstract No. 753.536 Chem. Abstr. 108: 212395f

Recent Contributions of Kolbe Electrolysis to Organic Synthesis

Hans-Jürgen Schäfer

Organisch-Chemisches Institut der Universität, Corrensstraße 40, D-4400 Münster, FRG

1 Introduction and Historical Background 92
2 Reaction Conditions for Kolbe-Electrolysis 93
3 Mechanism of Kolbe-Electrolysis 96
4 Symmetrical Coupling of Carboxylic Acids 99
5 Cross-Coupling of Different Carboxylates 104
6 Addition of Kolbe-Radicals to Double Bonds 110
7 Non-Kolbe Electrolysis to Carbenium Ions as Intermediates 115
8 Non-Kolbe Electrolysis of Carboxylic Acids to Ethers, Esters and Alcohols . . 117
9 Non-Kolbe Electrolysis of Carboxylic Acids to Acetamides 124
10 Conversion of Carboxylic Acids into Olefins by Non-Kolbe Electrolysis . . . 126
11 Rearrangement of Intermediate Carbocations in Non-Kolbe Electrolysis . . . 133
12 Fragmentation of Carboxylic Acids by Non-Kolbe Electrolysis 137
13 The Pseudo-Kolbe Reaction 138
14 The Photo-Kolbe Reaction 140
15 Anodic Oxidation of Carboxylic Acids Without Decarboxylation 141
16 Conclusions 142
17 Acknowledgement 142
18 References 143

Carboxylic acids can be decarboxylated by anodic oxidation to radicals (Kolbe-electrolysis) and/or carbocations (non-Kolbe electrolysis). The procedure and necessary equipment is simple, a scale-up easy, the choice of carboxylic acids wide, the selectivity towards radicals or carbocations can be controlled by reaction conditions and the structure of the carboxylic acid, the yields are in general good. The radical pathway can be used for the preparation of e.g. 1,n-diesters, pheromones, or rare fatty acids. Electrolysis in the presence of olefins affords additive dimers and monomers or by intramolecular addition five membered carbocycles and heterocycles. By non-Kolbe electrolysis carboxylic acids can be converted into ethers, acetals, olefins or acetamides. Rearrangements and fragmentations lead to stereospecifically substituted cyclopentanoids and one- or four-carbon ring extensions.

Hans-Jürgen Schäfer

1 Introduction and Historical Background

Kolbe electrolysis is a powerful method of generating radicals for synthetic applications. These radicals can combine to symmetrical dimers (chap 4), to unsymmetrical coupling products (chap 5), or can be added to double bonds (chap 6) (Eq. 1, path a). The reaction is performed in the laboratory and in the technical scale. Depending on the reaction conditions (electrode material, pH of the electrolyte, current density, additives) and structural parameters of the carboxylates, the intermediate radical can be further oxidized to a carbocation (Eq. 1, path b). The cation can rearrange, undergo fragmentation and subsequently solvolyse or eliminate to products. This path is frequently called non-Kolbe electrolysis. In this way radical and carbenium-ion derived products can be obtained from a wide variety of carboxylic acids.

$$R-CO_2^{\ominus} \xrightarrow[-CO_2]{-e} R \cdot \begin{array}{c}(a) \\ \nearrow \\ (a) \\ \longrightarrow \\ \searrow Y \end{array} \begin{array}{c} R-R \\ R-\!\!\!\!\begin{array}{c}Y\\|\\|\end{array}\!\!\!\!-R \end{array} \qquad (1)$$

$$-e \downarrow (b)$$

$$R^{\oplus} \longrightarrow \text{ester, ether, olefin, amide}$$

Faraday, in 1834, was the first to encounter Kolbe-electrolysis, when he studied the electrolysis of an aqueous acetate solution [1]. However, it was Kolbe, in 1849, who recognized the reaction and applied it to the synthesis of a number of hydrocarbons [2]. Thereby the name of the reaction originated. Later on Wurtz demonstrated that unsymmetrical coupling products could be prepared by coelectrolysis of two different alkanoates [3]. Difficulties in the coupling of dicarboxylic acids were overcome by Crum-Brown and Walker, when they electrolysed the half esters of the diacids instead [4]. This way a simple route to useful long chain 1,n-dicarboxylic acids was developed. In some cases the Kolbe dimerization failed and alkenes, alcohols or esters became the main products. The formation of alcohols by anodic oxidation of carboxylates in water was called the Hofer-Moest reaction [5]. Further applications and limitations were afterwards found by Fichter [6]. Weedon extensively applied the Kolbe reaction to the synthesis of rare fatty acids and similar natural products [7]. Later on key features of the mechanism were worked out by Eberson [8] and Utley [9] from the point of view of organic chemists and by Conway [10] from the point of view of a physical chemist. In Germany [11], Russia [12], and Japan [13] Kolbe electrolysis of adipic halfesters has been scaled up to a technical process.

The large amount of literature in connexion with the Kolbe and non-Kolbe electrolysis is covered in a number of reviews [7, 10, 16, 17, 22, 23, 24, 26, 27] and chapters in books [6, 9, 14, 15, 18, 19, 20, 21, 25].

2 Reaction Conditions for Kolbe Electrolysis

The yield and selectivity of Kolbe electrolysis is determined by the reaction conditions and the structure of the carboxylate. The latter subject is treated in chaps 3, 4. Experimental factors that influence the outcome of the Kolbe electrolysis are the current density, the temperature, the pH, additives, the solvent, and the electrode material.

High current densities and high carboxylate concentrations favor the formation of dimers. This is due to a high radical concentration at the electrode surface that promotes dimerization. Furthermore, at high current densities the so called critical potential of about 2.4 V (vs NHE) is reached [28] above which Kolbe dimerization proceeds smoothly. At this potential in aqueous solution and nonaqueous solvents the oxygen evolution and solvent oxidation is effectively suppressed [10, 29, 30]. The critical potentials for different carboxylates have been compiled [29], they are in the order of 2.1 to 2.8 V (vs NHE), whereby the structure of the carboxylate and its critical potential do not obviously correlate. There is, however, no need for potential control in Kolbe electrolysis as the critical potential is already exceeded at 1 to 10 mA/cm^2. This is much below the usually applied current density, which should be as high as possible, normally equal or greater than 250 mA/cm^2. It is assumed that at the critical potential the carboxylate forms a layer at the electrode surface, whereby the solvent is desorbed, whilst below the critical potential the solvent is preferentially oxidized.

In the anodic decarboxylation of phenylacetic acid benzaldehyde is the major product (80%) at low current density (< 3.2 mA/cm^2). Its formation is supposed to occur by reaction of the intermediate benzyl radical with oxygen, which is possibly simultaneously generated at the anode [31].

A neutral, or even better a weakly acidic medium seems to be preferable for the Kolbe reaction. This is achieved by neutralizing the carboxylic acid to an extent of 2 to 5%, in some cases up to 30%, by an alkali metal hydroxide or alkoxide. The concentration of carboxylate remains constant during the electrolysis. When the carboxylate is consumed at the anode, base is continuously formed at the cathode and this way the carboxylate is regenerated from the acid. The endpoint of the electrolysis is indicated by a change of the electrolyte to an alkaline pH. This procedure is denoted as the salt-deficit method. The degree of neutralization can be as low as 0.5%. In spite of the low conductivity in this case, a low cell voltage can be maintained by a small electrode distance (0.1 to 2 mm) and vibrating electrodes [32, 33] or capillary gap electrodes [34, 35]. Sometimes it is necessary to convert the carboxylic acid totally into the carboxylate to improve its solubility in aqueous solution or to suppress discrimination in mixed Kolbe electrolysis due to different acidities of the carboxylic acids. In these cases a mercury cathode is used in the Dinh-Nguyen cell [36]. Here the electrolyte remains neutral, because alkali metals are discharged and bound as amalgams.

In aqueous solution an elevated pressure favors the Kolbe-coupling against non-Kolbe products [37]. A possible explanation is that high pressure aids the formation of a lipophilic medium at the electrode surface that prevents the adsorption of water and thus disfavours the formation of carbenium ions.

Temperature has some effect on Kolbe electrolysis. Higher temperatures seem to support disproportionation against the coupling reaction and intramolecular additions to double bonds against a competing intermolecular coupling (chap 6). The unwanted conversion of the carboxylic acid into its methyl ester, that rivals to some extent with the decarboxylation, when methanol is used as solvent, can become the main reaction at higher temperatures, e.g. at about 65 °C with 11-bromoundecanoate [38].

Additives can strongly influence the Kolbe-reaction. Foreign anions should be definitively excluded, because they seem to disturb the formation of the necessary carboxylate layer at the anode. Their negative effect increases with the charge of the anion. While fluoride or formate inhibit, in low concentration, only slightly the ethane formation in the electrolysis of acetate, sulfate hinders the Kolbe reaction totally at concentrations as low as 10^{-3} mol/l [39]; even more pronounced, is the effect of $Fe(CN)_6^{3-}$ [22, 40]. In the anodic decarboxylation of phenylacetate the dimerization can be totally supressed in favor of the non-Kolbe product, benzylmethylether, when the electrolyte is more than 5.18 mM in $NaClO_4$ [31]. The inhibition by Cl^-, NO_3^-, NCS^-, Br^-, I^- can be partially counteracted by the addition of cyclohexene or pyridine [41]. The inhibition is attributed to the competitive formation of complexes Pt.PtO-foreign anion$_{ads}$ and Pt.PtO—$(RCO_2^-)_{ads}$. The former complex does not inhibit the dehydrogenation of the solvent, e.g. MeOH to $H_2C=O$ and $C\equiv O$. Unsaturated hydrocarbons are effective, because they trap the halogen from the anode surface. The shift from the radical to the carbenium ion pathway by perchlorate can be explained as due to blocking part of the anode surface by perchlorate adsorption, which lowers the radical concentration at the electrode. This disfavours the bimolecular radical dimerization and aids a second electron transfer to form the carbenium ion.

Foreign cations can increasingly lower the yield in the order Fe^{2+}, Co^{2+} < Ca^{2+} < Mn^{2+} < Pb^{2+} [22]. This is possibly due to the formation of oxide layers at the anode [42]. Alkali and alkaline earth metal ions, alkylammonium ions and also zinc or nickel cations do not effect the Kolbe reaction [40] and are therefore the counterions of choice in preparative applications. Methanol is the best suited solvent for Kolbe electrolysis [7, 43]. Its oxidation is extensively inhibited by the formation of the carboxylate layer. The following electrolytes with methanol as solvent have been used: MeOH-sodium carboxylate [44], MeOH—MeONa [45, 46], MeOH—NaOH [47], MeOH—Et_3N-pyridine [48]. The yield of the Kolbe dimer decreases in media that contain more than 4% water.

In aqueous solution especially, the current yield is distinctly lower; furthermore, solubility problems can occur when the salt-deficit method is used. In aqueous solution, α-amino- or α-phenyl substituted carboxylates lead mainly to decomposition products, whilst in dry methanol or methanol-pyridine, coupling products were obtained with α-phenyl- and α-acetylaminocarboxylates [49].

Dimethylformamide is also a suitable solvent [50], it has, however, the disadvantage of being oxidized at fairly low potentials to *N*-acyloxy-*N*-methyl formamide [51]. The influence of the composition of the ternary system water/methanol/dimethylformamide on the material and current yield has been systematically studied in the electrolysis of ω-acetoxy or -acetamido substituted carboxylates [32]. Acetonitrile can also be used, when some water is added [52]. The influence of various solvents on the ratio of Kolbe to non-Kolbe products is shown in Table 1 [53].

Other electrolytes, that have been used, are: dimethylsulfoxide-NaH [54], glycol methylether-water (3:7, v, v) [55] or acetic acid [22].

Table 1. [53] Anodic decarboxylation[a] of cyclohexanecarboxylic acid in different solvents

Solvent	Bicyclohexyl:Carbeniumion product[b]
H_2O-MeOH (30% v/v)[c]	0.97
MeOH[c]	1.76
MeCN[d]	2.26
$HCONMe_2$[d]	4.32

[a] Acid (0.78 M), neutralized to 25%; Pt-anode, 0.25 A/cm^2
[b] Alcohol, ester, ether and acetamide according to solvent
[c] Sodium salt
[d] Tetrabutylammonium-salt

As anode material, smooth platinum in the form of a foil or net seems to be most universally applicable [32, 33]. In nonaqueous media, platinized titanium, gold, and nonporous graphite can also be used [56]. PbO_2-, MnO_2- or Fe_3O_4-anodes do not lead to Kolbe-dimers [57], except for PbO_2 in acetic acid [58].

To keep the consumption of the valuable platinum low, thin foils have been glued to a graphite support [34] or thin-layers of platinum have been sputtered on to a glass base [59]. Platinum can be used in a particle electrode by plating silica gel with platinum [60], or in a solid polymer electrolyte where platinum is incorporated into a nafion ion exchange membrane [61].

Electrolysis of triethylammonium heptanoate gave, in protic solvents (MeOH, EtOH, H_2O), dodecane as major the product at vitreous carbon (24–53%) or baked carbon (30–39%), compared to 45–62% at platinum; with graphite as anode, esters (non-Kolbe products) were the major products. Additionally, large differences in the product ratios were found by using face or edge surfaces of pyrolytic graphite [48]. In acetonitrile, longer fatty acids form dimers also at graphite anodes. Is has been argued that this is possibly due to stacking of the alkyl chain at the electrode surface [52]. In acetonitrile, glassy carbon behaves like platinum in the anodic oxidation of acetate [62]. In non-aqueous media, ruthenium dioxide-titanium anodes have been successfully applied to Kolbe dimerizations [63].

The nature of the cathode material is not critical in the Kolbe reaction. The reduction of protons from the carboxylic acid is the main process, so that the electrolysis can normally be conducted in an undivided cell. For substrates with double or triple bonds, however, a platinum cathode should be avoided, as cathodic hydrogenation can occur there. A steel cathode should be used, instead.

In summary the following general experimental conditions should be applied for a successful dimerization of carboxylic acids: An undivided beaker type cell can be used equipped with a smooth platinum anode and a platinum, steel or nickel cathode in close distance; a current density of 0.25 A/cm^2 or higher should be provided by a regulated power supply, a slightly acidic or neutral electrolyte, preferable methanol as solvent and a cooling device to maintain temperatures between 10 to 45 °C should be employed. With this simple procedure and equipment yields of coupling product as high as 90% can be obtained, provided the intermediate radical is not easily further oxidized (see chap 7).

3 Mechanism of Kolbe Electrolysis

In the 1930s, very different mechanisms for Kolbe electrolysis were proposed. The discharged ion or free-radical theory was suggested by Crum, Brown and Walker [64], it assumes acyloxy radicals as intermediates. This proposal comes closest to present mechanistic ideas. According to Glasstone's and Hickling's hydrogen peroxide theory [65], Kolbe dimers and Hofer-Moest products (alcohols, esters) are formed by the action of OH· radicals and H_2O_2. The acyl peroxide theory by Schall [66] and Fichter [6] proposes that the acyloxy radicals formed by discharge of the carboxylates couple to diacyl peroxides, that decompose. These early mechanistic ideas are covered in refs [67, 68]. More recent mechanistic concepts explain the experimental facts better. However, there is still some dispute between physical and organic chemists on some reaction steps. The recent mechanistic proposals have been critically discussed in refs. [8, 69–72] and from the physical chemists points of view in ref. [10]. The following general scheme is assumed (Eq. 2)

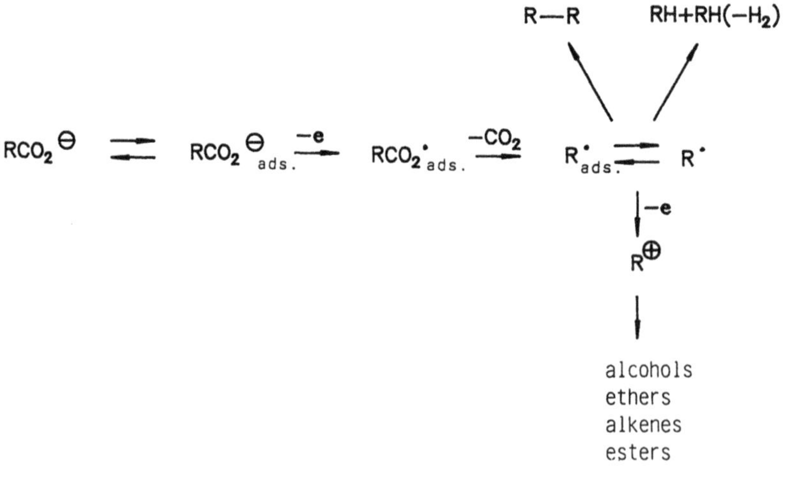

$$RCO_2^{\ominus} \rightleftarrows RCO_{2\,ads.}^{\ominus} \xrightarrow{-e} RCO_{2\,ads.}^{\cdot} \xrightarrow{-CO_2} R_{ads.}^{\cdot} \rightleftarrows R^{\cdot} \nearrow^{R-R}_{RH+RH(-H_2)} \quad (2)$$

$$\downarrow -e$$
$$R^{\oplus}$$
$$\downarrow$$
alcohols
ethers
alkenes
esters

The competing pathways to radical or carbenium ion derived products are determined, apart from experimental factors (see chap. 2), by the ionization potential of the radical. From product ratios and ionization potentials of the intermediate radicals, the conclusion could be drawn that such radicals with ionization potentials above 8 eV lead preferentially to coupling products, whilst those with ionization potentials below 8 eV are further oxidized to carbenium ions [8c].

Pulsed current experiments of aqueous acetate solutions indicate that at least in aqueous solution a platinum oxide layer seems to be prerequisite for the decarboxylation to occur. Only at longer pulse durations ($> 10^{-3}$ s) is ethane produced [73, 74]. These are times known to be necessary for the formation of an oxide film. At a shorter pulse length ($< 10^{-4}$ s) acetate is completely oxidized to carbon dioxide and water possibly at a bare platinum surface [75]. The potentiodynamic response in the electrolysis of potassium acetate in aqueous solution also points to an oxide layer, whose average thickness is probably equivalent to three monolayers [76]. Other results, however, indicate that platinum oxides formed in the presence of water significantly inhibit the Kolbe reaction. This is assumed from a comparison of reaction rates in anhydrous and aqueous solutions [77] and the fact that the highest yield of dimers are usually obtained in nonaqueous media [78].

It has been shown by employing the radioactive tracer method with ^{14}C-labeled carboxylic acids [79] and with rotating disc electrode experiments [80] that carboxylates are adsorbed at the anode surface.

Current potential curves exhibit a critical potential at 2.1 to 2.4 V versus NHE. Below this potential in aqueous solution, oxygen is formed preferentially [81], while above it the discharge of oxygen is inhibited and Kolbe dimers are obtained [81–82]. At this critical potential the coverage of the electrode with acetate increases sharply [83]. From an anomalous Tafel slope, from the decrease of the differential capacity in this voltage range [83] and from other electroanalytical data [84] it is deduced that above the critical potential a rigid layer of adsorbed acyloxy or alkyl radicals is formed on top of the metal [8c, 78] or metal oxide [68], which inhibits the oxygen evolution or solvent oxidation and promotes Kolbe electrolysis [8a, b, 10, 76]. Recent results on the capacitative behaviour of sodium acetate in an aqueous acetic acid solution at platinum indicate that the double layer capacitance is due to adsorption of methyl radicals, while the coverage by acetate radicals is insignificant [85]. The different stages of oxide formation and adsorption of intermediates have been observed by modulated specular reflectance spectroscopy applied to the Kolbe reaction of ethyl malonate in aqueous solution [86]. By polaromicrotribometry is has been shown that at the critical potential a hydrophobic polymer film is formed that apparently inhibits the oxygen discharge [87].

For the tetrabutylammonium salts of substituted acetate the quarter wave potentials have been determined by chronopotentiometry in acetonitrile. The ease of oxidation, as reflected in the $E_{1/4}$-values, decreases with increasing strength of the acid [88].

The current-potential relationship indicates that the rate determining step for the Kolbe reaction in aqueous solution is most probably an irreversible 1 e-transfer to the carboxylate with simultaneous bond breaking leading to the alkyl radical and carbon dioxide [8]. However, also other rate determining steps have been proposed [10]. When the acyloxy radical is assumed as intermediate it would be very shortlived and decompose with a half life of $\tau \approx 10^{-8}$ to carbon dioxide and an alkyl radical [89]. From the thermochemical data it has been concluded that the rate of carbon dioxide elimination effects the product distribution. Olefin formation is assumed to be due to reaction of the carboxylate radical with the alkyl radical and the higher olefin ratio for propionate and butyrate is argued to be the result of the slower decarboxylation of these carboxylates [90].

Whilst electroanalytical data have been interpreted in terms of a mechanism involving the coupling of adsorbed radicals [10, 91] the stereochemistry of the products and the regioselectivity of the coupling reaction indicates that adsorption of saturated alkyl radicals seems to be relatively unimportant. Support that non or weakly adsorbed radicals combine, comes from the coupling products of cyanoalkyl radicals **1** [8]. These radicals with two reaction sites produce the same amount of carbon to carbon and carbon to nitrogen coupling product, when generated by photolytic or thermal decomposition of an azonitrile, by persulfate oxidation or by electrolysis of α-cyanoacetic acid in methanol, acetonitrile or dimethylformamide. In sharp contrast, photolysis of the same azonitrile adsorbed on silica produces only the carbon to carbon coupling product. Similarly the addition of radical **4** to butadiene affords the same ratio of 1,1-, 1,3- and 3,3-additive dimers, irrespective of whether the radical was generated by anodic or persulfate oxidation of methyl malonate **2** or reductive fragmentation of the hydroperoxide **3** (Eq. 3) [92].

Carboxylates, which are chiral in the α-position totally lose their optical activity in mixed Kolbe electrolyses [93, 94]. This racemization supports either a free radical or its fast dynamic desorption-adsorption at the electrode. A clearer distinction can be made by looking at the diastereoselectivity of the coupling reaction. Adsorbed radicals should be stabilized and thus react via a more product like transition state that reflects different stabilities of the products. The free radical on the other side should have a more educt like transition state and should therefore couple more randomly. This has been tested with the carboxylic acids 5 to 9. The saturated carboxylic acids 5 and 6 coupled randomly, which indicated that these radicals are not significantly adsorbed. The unsaturated carboxylic acids 7 to 9 did not dimerize quite statistically, which agrees with the fact that unsaturation favors adsorption [95, 96].

The intermediate radical (Eq. 2) can be further oxidized to a carbenium ion that undergoes solvolysis, rearrangement and elimination (see chaps 8–12). This oxidation is strongly influenced by experimental factors as for example the presence of foreign anions, or the electrode material (see chap. 2). If one assumes, that the radical is oxidized from an adsorbed state the strong increase of the benzyl cation formation in the phenylacetatic acid oxidation by added perchlorate can be explained by the competitive preferential adsorption of the perchlorate anion [31]. With nuclear substituted phenyl acetates the portion of radical coupling products increases with increasing electron withdrawing ability of the substituents, for example with *p-tert*-butyl it is 13% (ionization potential of the benzyl radial = 7.4 eV) and for unsubstituted phenyl it is 76% (IP = 7.76 eV). For the oxidation of *p*-methoxyphenylacetic acid a pseudo-Kolbe reaction (see chap. 13) is assumed, which means an electron transfer from the aromatic nucleus [31]. Besides its dimerization or addition to double bonds (chap. 6) the radical can disproportionate or react with oxygen [31, 97]. The ratio of disproportionation to dimerization seems to be higher in Kolbe electrolysis than in an homogeneous reaction [98].

4 Symmetrical Coupling of Carboxylic Acids

Two equal carboxylates can be coupled to symmetrical dimers (Eq. 4). In spite of the high anode potential, that is necessary for Kolbe electrolysis, a fair number of

$$R^3-\underset{\underset{R^2}{|}}{\overset{\overset{R^1}{|}}{C}}-CO_2^{\ominus} \quad \underset{-CO_2}{\overset{-e}{\rightleftarrows}} \quad R^3-\underset{\underset{R^2}{|}}{\overset{\overset{R^1}{|}}{C}}-\underset{\underset{R^2}{|}}{\overset{\overset{R^1}{|}}{C}}-R^3 \quad (4)$$

R^1, R^2, R^3: H, Alkyl, Arylalkyl

$R^1, R^2 = H$, R^3: CO_2Me, $(CH_2)_nX$ (X = COR, CO_2R, n⩾1,

X = OAc, NHAc, Hal, n⩾4)

substituents can be present in the carboxylic acid. As the reaction involves radicals as reactive intermediates, polar substituents can be handled without protection. This saves additional steps for protection and deprotection, that are often necessary in polar reactions, where strong bases, nucleophiles or electrophiles are used as reagents. Furthermore carboxylic acids are readily available in a wide variety of structures. These advantages make Kolbe electrolysis a method of choice for the construction of symmetrical compounds.

Table 2. Symmetrical coupling of selected carboxylic acids

No.	Substituents in $R^1R^2R^3C\text{-}CO_2^-$			Yield[a] of coupling product (%)	Ref.
	R^1	R^2	R^3		
1	H	H	H	93	7, 9
2	Alkyl (C_1–C_{17})	H	H	30–90	7, 9
3	Alkyl	Alkyl	H	0–26	9
4	$RO_2C\text{-}(CH_2)_n$ n = 3–15, R = Me, Et	H	H	40–95	7
5	Alkyl	EtO_2C	H	20–85	99, 100
6	Alkyl	Alkyl	EtO_2C	15–35	99, 101
7	Alkyl	CN	H, Alkyl	30–60	102
8	Alkyl	$CONH_2$	H, Alkyl	5–55	103
9	Alkyl-CO	Alkyl, H	Alkyl, H	32–40	104
10	$CH_3CO(CH_2)_n$	H	H		
	n = 1			70	105
	n = 5			63	106
11	F, Br, Cl, I, OH, NH_2	H	H	0	7, 9, 107
12a	F	F	F	93	84b, 108
12b	F	F	Cl	43	109
13	$X(CH_2)_n$	H	H		
	X: F, n > 3			45–70	110
	X: Cl, n: 1, 3, 4, 5, 7, 8, 9, 11			40–80	110, 111, 112
	X: Br, n: 4–11, 14			54–71	38a, 110
14	$RO_2C(CF_2)_n$ n = 0–3, R = Me, Et	F	F	high yield	113
15	$H(CF_2)_n$ n = 3, 5	F	F	40–86	114
16	$CF_3(C_2F_5)_2C$	H	H	75	115
17	$CH_3CO_2(CH_2)_n$ n: 2–5	H	H	70–83	32
18	$CH_3CONH\text{-}(CH_2)_4$	H	H	24–39	32, 116
19	Phthaloyl–N–CHCH$_2$ | CO_2H	H	H		32, 117
20	X–C$_6$H$_4$ X: p-OMe to X: F$_5$	H	H	>1–74	31
21	Ph	Ph	H	25	118
22	mesityl	H	H	45	118
	2-methylbiphenyl	H	H	39	119

[a] Electrolyses are normally performed at a platinum anode in water, methanol, ethanol or ethanol/water

A great number of Kolbe dimerizations have been tabulated in refs. [9, 17–19]. Here no comprehensive coverage is intended, but to demonstrate with selected examples the range and limitations of Kolbe dimerization. In the following discussion and in Table 2 the carboxylates are arranged according to their functional groups in the order alkyl-, ester-, keto-, halo- and olefinic substituents.

In general, only the kind of substituent in the α-position is critical for the yield of the coupling product. Electron donating groups (more than one alkyl group, phenyl, vinyl, halo or amino substituents) more or less shift the reaction towards products that originate from carbenium ions (non-Kolbe products, see chap. 7). Electron attracting groups (cyano, ester or carbonyl substituents) or hydrogen, on the other side, favor the radical dimerization.

The reaction conditions, normally applied, are those described in chap. 2 for the radical pathway. These are a platinum anode, high current densities, no additives and a slightly acidic medium. However, the dimerizations shown in Table 2, No. 2, also gave in some cases good yields at a carbon anode in acetonitrile-water [52] or at a baked carbon anode in methanol [48]. With propionic and butyric acid an unusually high portion of alkene is formed at the cost of the dimer.

The coupling of carboxylic acids has been profitably used in natural product synthesis. Kolbe electrolysis of **10** is part of a (+)-α-onocerin synthesis [120], the dimerization of **11** leads to a pentacyclosqualene [121], the electrolysis of **12** afforded a dimer with two quaternary carbon atoms [122], and 2,6,10,15,19,23-hexamethyltetracontane has been synthesized from **13** [123].

For the preparation of long-chain alkanes it proved useful to extract the electrolyte continously with a high boiling point nonpolar solvent. This way the alkanes: $C_{34}H_{70}$, $C_{38}H_{78}$ and $C_{42}H_{86}$ were prepared with high yields (70, 75 and 87%, respectively) [124].

Some cyclopropylcarboxylic acids, namely **14** [125] and **15** [46] could be coupled to bicyclopropyl compounds, others led to allylic compounds via ring opening of an intermediate carbenium ion (see chap. 7). Tertiary alkanoates yield predominantly non-Kolbe products (see chap. 8).

The Brown-Walker version of Kolbe electrolysis namely the dimerization of half-esters to diesters of diacids, is of industrial interest, because using this method, sebacates can be prepared from adipic half esters [126]. This process appears to have reached the pilot stage in Germany [11], Japan [13], and Russia [12]. In Russia, the behaviour of other half esters has also been studied in detail [12]. From China, too, activities concerning this coupling reaction have recently been reported [127]. Yields as high as 93% have been obtained [128]. Bis(2-ethylhexyl)sebacate, a plasticizer, has been prepared from the corresponding adipate in 85% yield [128]. In sebacate production, platinum can be replaced by a special carbon anode, which gives nearly the same yields as platinum and is highly resistant to the electrolyte [129].

Efficient syntheses of substituted succinic acids (Table 2, Nos. 5, 6) have been developed in the past; a more recent application is the coupling of **16** as part of a semibullvalene synthesis [130].

α,ω-Dicarboxylated polyethylene or polydifluormethylene is reported to be obtained by electrolysis of azelaic acid [131] or perfluoroglutaric acid [132].

16 28% (MeOH,Pt)

17 65%

18 67%

Whilst ketocarboxylic acids can be dimerized satisfactorily (Table 2, No. 9, 10), the corresponding aldehydes couple poorly. Good yields can be obtained in these cases when the acetals, e.g. **17** [133], **18** [134], are electrolyzed instead.

Carboxylic acids with a halide, hydroxy or amino group in the α-position form no dimers (Table 2, No. 11), except when two or three fluorine atoms are present there (Table 2, No. 12). A large amount of work has been devoted to the coupling of fluorocarboxylic acids (Table 2, Nos. 13–16) due to interesting properties of the produced fluorohydrocarbons. By statistical analysis optimal conditions for Kolbe electrolysis of perfluorinated acids have been calculated [135].

Hydroxy- and amino carboxylic acids can be dimerized in good to moderate yields, when the substituents are not in the α- or β-position and when they are additionally protected against oxidation by acylation (Table 2, No. 17–19). 2-Alkenoic acids cannot be dimerized but lead to more or less extensive passivation of the anode due to the formation of polymer films [136]. 3- and 4-Alkenoic acids give moderate yields when they are neutralized with Bu_3N or Et_3N [136]. 3-Alkenoic acids with the structure of unsymmetrical half esters, derived from Stobbe condensations, afford dimers in 14–30% yield, major side products are ethers arising from non-Kolbe electrolysis [137]. 3-Alkenoic acids dimerize to a mixture of three 1,5-dienes (Eq. 5), that arise by 1,1'-, 1,3'- und 3,3'-coupling of the intermediate allyl radical [138]. When the 3-position of the allyl radical is increasingly sterically shielded, the ratio of 3-coupling decreases. The relative amount of the 1,1'-dimer thus can vary from 52 to 76% (Table 3). The configuration of non-terminal double bonds is retained to a high degree (~90%) [138a].

$$(5)$$

Table 3. Yields in Kolbe dimerization of 3-alkenoic acids [138b]

Carboxylic acid[a]	Product distribution (%)			Yield (%)
	1,1'	1,3'	3,3'	
$CH_3-(CH_2)_7-CH=CH-CH_2-CO_2^-$	52	39	9	67
$(CH_3)_2-CH-CH=CH-CH_2-CO_2^-$	59	41	—	79
$(CH_3)_3-C-CH=CH-CH_2-CO_2^-$	60	40	—	15
⌬—$CH_2-CO_2^\ominus$	60	40	—	45
$(CH_3)_2-CH=CH-CH_2-CO_2^-$	65	—	36	42
⌬=$CH-CH_2-CO_2^\ominus$	76	24	—	29

[a] 10–150 mmol were dissolved in 25–100 ml methanol, neutralized to 8–50% with triethylamine and electrolyzed in an undivided cell at platinum electrodes at 400–800 mA/cm^2 with change of polarity of the electrodes until pH = 8 was reached

$$(6)$$

With 6-alkenoic acids the intermediate radical partially cyclizes to a cyclopentyl-methyl radical in a 5-*exo-trig* cyclization [139] (Eq. 6) [138a, 140] (see also chap. 6). To prevent double bond migration with enoic acids the electrolyte has to be hindered to become alkaline by using a mercury cathode. Z-4-Enoic acids partially isomerize to *E*-configurated products. Results from methyl and deuterium labelled carboxylic acids support an isomerization by way of a reversible ring closure to cyclopropyl-carbinyl radicals. The double bonds of Z-N-enoic acids with N > 5 fully retain their configuration [140].

The acid **19** has been dimerized, although in low yield, in the course of a perhydro-phenanthrene synthesis [141]. When the oxidation potential of the double bond is sufficiently lowered by alkyl substituents, lactone formation by oxidation of the couble bond rather than of the carboxyl group occurs (Eq. 7) [142] (see also chap. 15).

Other alkenoic acids that have been dimerized with retention of configuration at the double bond are oleic acid (23% dimer yield), elaidic acid (44%) [143], and erucic acid (40%) [144].

Substituted phenylacetic acids form Kolbe dimers when the phenyl substituents are hydrogen or are electron attracting (Table 2, Nos. 20–23); they yield methyl ethers (non-Kolbe products), when the substituents are electron donating (see also chap. 8). Benzoic acid does not decarboxylate to diphenyl. Here the aromatic nucleus is rather oxidized to a radical cation, that undergoes aromatic substitution with the solvent [145].

5 Cross-Coupling of Different Carboxylates

Cross coupling of two different carboxylates (= mixed Kolbe electrolysis) is a method for synthesizing unsymmetrical compounds (Eq. 8). As, however, the intermediate radicals combine statistically, the mixed coupling product

$$4\ ^1RCO_2H + 4\ ^2RCO_2H \xrightarrow[-CO_2]{-e} 4\ ^1R + 4\ ^2R \longrightarrow$$

$$^1R\text{---}^1R + 2\ ^1R\text{---}^2R + ^2R\text{---}^2R$$

(8)

is always accompanied by two symmetrical dimers as major side products. To make this coupling more attractive for synthesis the less costly acid is taken in excess. This way the number of major products is diminished to two, which facilitates the isolation of the mixed dimer. Furthermore the more costly acid is incorporated to a large extent into the mixed dimer. Some calculated yields are listed in Table 4. Some experimental yields obtained for cross-coupling of methyl adipate with hexanoate are compiled in Table 5.

When one of the two acids is used in excess and the pk_a-values of the two acids differ strongly, the salt deficit method should be used with caution. Formic acid, acetic acid, propionic acid, and trifluoroacetic acid have been electrolyzed competitively in mixtures of pairs. Formic acid and trifluoroacetic acid are comparable in case of electrolysis, both are more readily electrolyzed than acetic and propionic acids. Deviations are rationalized on the basis of differences in ionization [147]. It might be useful in such cases to neutralize both acids completely. Sometimes one of the two acids, although being the minor component, is more favorably oxidized possibly due to preferential adsorption or its higher acidity [148]. In this case the continuous addition of the more acidic acid to an excess of the weaker acid may lead to successful cross-coupling [149]. The chain length of the two acids should be chosen in such a

Table 4. Calculated theoretical yields for mixed coupling products ^1R–^2R

$^1RCO_2H$	:	$^2RCO_2H$	Yield of ^1R–^2R (%)[a]
1	:	1	50
1	:	2	66
1	:	4	80
1	:	6	86
1	:	10	91

[a] Calculated according to %-Yield = $n \cdot 100/(1 + n)$ with 1:n as ratio of $^1RCO_2H$:$^2RCO_2H$

Table 5. Yield of methyl decanoate in the cross-coupling of methyl adipate and hexanoate [146]

Ratio of $HO_2C(CH_2)_4CO_2CH_3$	:	$CH_3(CH_2)_4CO_2H$	%-Yield of $MeO_2C(CH_2)_8CH_3$ (%)	
			in MeOH	in H_2O
1	:	1	36	12
1	:	2	49	39
1	:	6	58	48

way that the symmetrical dimer formed in excess can be separated from the cross-coupling product either by distillation or crystallization. For smaller scale reactions the selective coupling of two different alkyl groups can be achieved by photolysis of unsymmetrical peroxides at low temperatures in the solid state [150].

Problems due to passivation that lead to an increase of the cell voltage or due to competition by non-Kolbe electrolysis [179] are often less pronounced in mixed coupling.

Despite of the disadvantage. that at least one symmetrical dimer is formed as a major side product, mixed Kolbe electrolysis has turned out to be a powerful synthetic method. It enables the efficient synthesis of rare fatty acids, pheromones, chiral building blocks or non proteinogenic amino acids. The starting compounds are either accessible from the large pool of fatty acids or can be easily prepared via the potent methodologies for the construction of carboxylic acids.

Another advantage of the synthesis by mixed Kolbe electrolysis is that polar groups in the carboxylic acid are tolerated in radical coupling. This makes additional protection-deprotection steps unneccessary, which are often needed in polar CC-bond forming reactions and can make these approaches less attractive in such cases.

Selected examples of the large number of compounds synthesized since the pioneering work of Weedon are subsequently arranged either in Tables 6, 7 or in formulas. At first the coupling products between unsubstituted alkyl groups and substituted

Table 6. Cross-coupling by Kolbe electrolysis of unsubstituted (A) with substituted carboxylic acids (B)

Carboxylic acids		Crosscoupling product (A–B) Yield (%)	Ref.
A–CO_2H	B–CO_2H		
$CH_3(CH_2)_6CO_2H$	$CH_3O_2C(CH_2)_4C\equiv C(CH_2)_4CO_2H$	23	151
$CH_3(CH_2)_{14}CO_2H$	$(CH_3)_2CHCO_2H$	40	152
$CH_3(CH_2)_5CO_2H$	(E)–$CH_3(CH_2)_7CH=CH(CH_2)_7CO_2H$	37	152
CD_3CO_2H	CF_3CO_2H	68	149
CD_3CO_2H	$CF_3CF_2CO_2H$	75	149
CH_3CO_2H	$HC\equiv C-(CH_2)_8CO_2H$	36	153
$CH_3CH_2CO_2H$	$C_6H_5-(CH_2)_2CO_2H$	44	154
$CH_3(CH_2)_3CO_2H$	$CH_3O_2C(CH_2)_7CO_2H$	69	144
$CH_3(CH_2)_5CO_2H$	$CH_3O_2C(CH_2)_7CO_2H$	61	144
$CH_3(CH_2)_2CO_2H$	$CH_3O_2CCH(OCOCH_3)CO_2H$	29	155
$CH_3(CH_2)_2CO_2H$	$CH_3CO(CH_2)_4CO_2H$	62	156
Methylhexadecanoic acid	$CH_3O_2C(CH_2)_4CO_2H$	42	157
$CH_3(CH_2)_8CO_2H$	$CH_3O_2C(CH_2)_4CO_2H$	38	146
$CH_3(CH_2)_2CO_2H$	$CH_3O_2CCH_2CH(CH_3)CH_2CO_2H$	30	158
⬠–CH_2CO_2H	$CH_3O_2C(CH_2)_{11}CO_2H$	30	159
$(CH_3)_3C(CH_2)_2CH(CH_3)-$ $-CH_2CO_2H$	$CH_3O_2CCH_2CH(C_2H_5)-$ $-CH_2CO_2H$	15	160
$CH_3(CH_2)_{10}CO_2H$	$CH_3O_2CCH(OCH_3)CH_2CO_2H$	21	161
$CH_3(CH_2)_7CH(CH_3)CH_2CO_2H$	$CH_3O_2C(CH_2)_7CO_2H$	30	162
$Me(CH_2)_mCO_2H$	$CH_3O_2C(CH_2)_nCO_2H$		
m = 8–16, 18	n = 4, 7, 8, 12–20	–	163
i-$C_4H_9CO_2H$	$(C_2H_5)_2C(CO_2Et)CO_2H$	38	164

alkyl groups are described, whereby the latter are ordered according to the presence of double and triple bonds, hydroxy groups, halides, carbonyl, carboxyl, ester, amino groups and other heterosubstituents. These examples are than followed by cross-coupling products between two substituted alkyl groups in the same order. In the formulas often the product of subsequent transformations is shown, the coupling site is marked by an arrow and the yield in mixed Kolbe electrolysis is indicated. Further examples are given in references [9, 23].

A large number of trialkylacetic acid esters have been prepared by mixed Kolbe electrolysis of ethyl malonates [164]. Crossed-coupling is also used for chain extension. Extension by two carbon atoms is achieved with benzyl succinates [153, 180–182], whereby the purification of the chain extended fatty acid is simplified by using the benzyl half ester [181a].

Extension by three carbon atoms is possible with methyl glutarate [183], by the isoprene unit with ethyl 3-methyl adipate [184], by four carbon atoms with methyl adipate [143], by five carbon atoms with methyl pimelate [185] and by six carbon atoms with methyl suberate [186]. A series of branched ω-fluorocarboxylic acids were prepared by cross-coupling with ω-fluorocarboxylic acids [187]. For further examples see Tables 6, 7.

Table 7. Cross-coupling by Kolbe electrolysis of substituted carboxylic acids (A) with substituted carboxylic acids (B)

Carboxylic acids		Crosscoupling product (A–B) Yield (%)	Ref.
A–CO_2H	B–CO_2H		
$CH_2=CH(CH_2)_2CO_2H$	$CH_3O_2C(CH_2)_{10}CO_2H$	68	165
$(CH_3)_2CHCH=CHCH_2CO_2H$	$CH_3O_2C(CH_2)_nCO_2H$ $n = 1, 2, 4$	good	166a
$(Z)-CH_3O_2C(CH_2)_2CH=CH(CH_2)_2CO_2H$	$CH_3(CH_2)_2CO_2H$	30	166b
$CH_2=CH(CH_2)_8CO_2H$	$CH_3O_2C(CH_2)_2CO_2H$	52	166c
$R-C\equiv C-(CH_2)_3CO_2H$ R: C_2H_5, C_3H_7, C_4H_9	$CH_3O_2C(CH_2)_nCO_2H$ $n = 2, 3, 4, 6$	45–59	167
$(Z)-CH_3(CH_2)_7CH=CH-(CH_2)_7CO_2H$	$CH_3O_2C(CH_2)_2CO_2H$	15–20	168
$CF_3CF_2OCF_2CO_2H$	$MeO_2C(CF_2)_nCO_2H$ $n = 1-3$	24–34	169
$CF_3OCF_2CF_2CO_2H$	$MeO_2C(CF_2)_nCO_2H$ $n = 1-3$	24–34	169
CF_3CO_2H	$CH_3O_2CCH_2CO_2H$	46	170
$Br(CH_2)_{10}CO_2H$	$CH_3O_2C(CH_2)_7CO_2H$	50	144
$Br(CH_2)_{10}CO_2H$	$EtO_2C(CH_2)_8CO_2H$	54	171
$Br(CH_2)_{10}CO_2H$	$MeO_2C(CH_2)_4CO_2H$	37	38b
$AcO(CH_2)_4CO_2H$	$EtO_2C(CH_2)_{10}CO_2H$	27–32	172, 178
$Me_3SiCH_2CO_2H$	$MeO_2C(CH_2)_nCO_2H$ $n = 4, 7$	71–76	173
$CH_3O_2C(CH_2)_{11}CO_2H$	$CH_3O_2C(CH_2)_4CO_2H$	—	174
$CH_3O_2C(CH_2)_{10}CO_2H$	$CH_3O_2C(CH_2)_4CO_2H$	—	175
$CH_3O_2C(CF_2)_nCO_2H$ $n = m = 1-4$	$CH_3O_2C(CF_2)_mCO_2H$	25–38	113
$CH_3CO(CH_2)_2CO_2H$	$CH_3CO(CH_2)_4CO_2H$	31	176
$CH_3CO(CH_2)_8CO_2H$	$CH_3O_2C(CH_2)_4CO_2H$	32	177

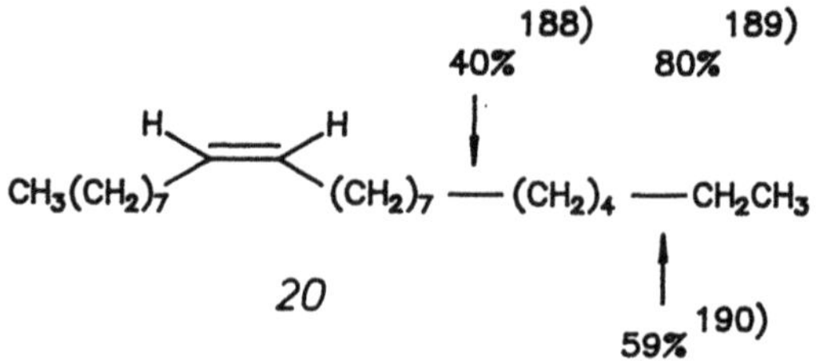

$$CH_3(CH_2)_7\overset{H}{\underset{}{\diagup}}=\overset{H}{\underset{}{\diagdown}}(CH_2)_7\xrightarrow[]{40\% \;\;188)} (CH_2)_4\xrightarrow[59\%\;190)]{80\%\;189)} CH_2CH_3$$

20

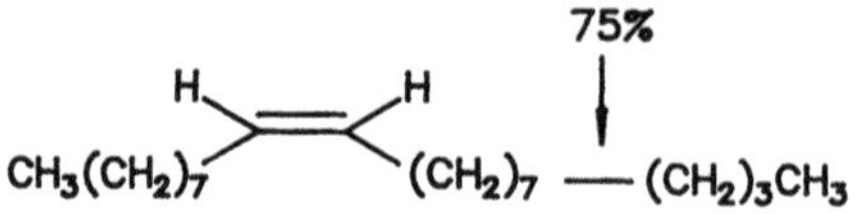

$$CH_3(CH_2)_7\overset{H}{\underset{}{\diagup}}=\overset{H}{\underset{}{\diagdown}}(CH_2)_7 \xrightarrow{75\%} (CH_2)_3CH_3$$

21

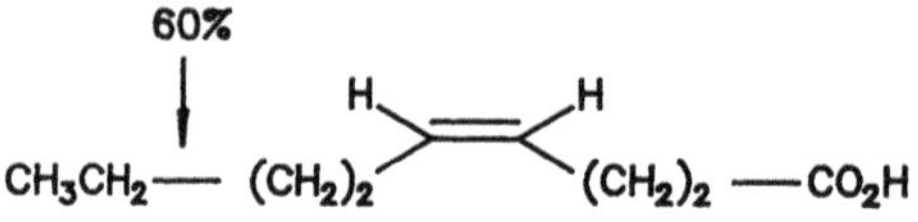

$$CH_3CH_2 \xrightarrow{60\%} (CH_2)_2 \overset{H}{\underset{}{\diagup}}=\overset{H}{\underset{}{\diagdown}} (CH_2)_2 - CO_2H$$

22

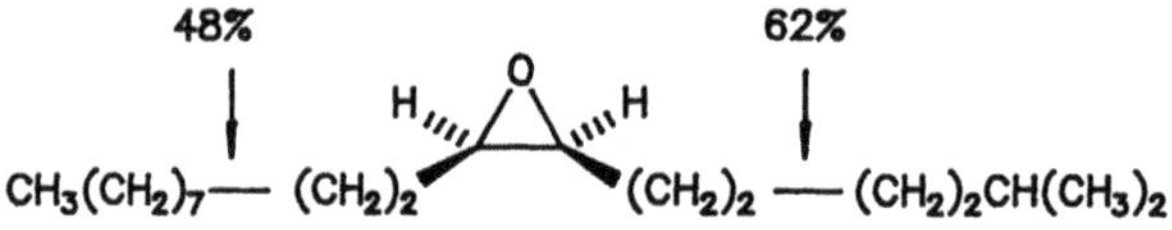

$$CH_3(CH_2)_7 \xrightarrow{48\%} (CH_2)_2 \overset{O}{\underset{}{\triangle}} (CH_2)_2 \xrightarrow{62\%} (CH_2)_2CH(CH_3)_2$$

23

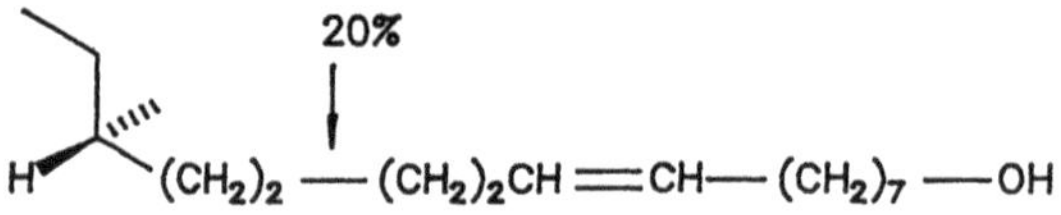

$$H\cdots\overset{Et}{\underset{}{C}}(CH_2)_2 \xrightarrow{20\%} (CH_2)_2CH=CH-(CH_2)_7-OH$$

24

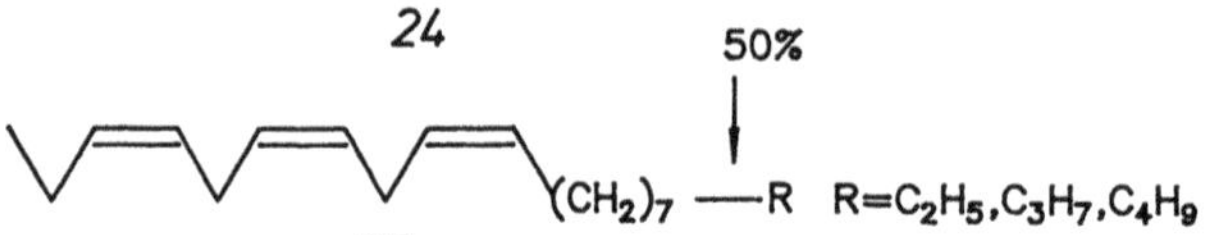

$$\diagup\!\!=\!\!\diagup\!\!=\!\!\diagup\!\!=\!\!\diagdown (CH_2)_7 \xrightarrow{50\%} R \quad R=C_2H_5, C_3H_7, C_4H_9$$

25

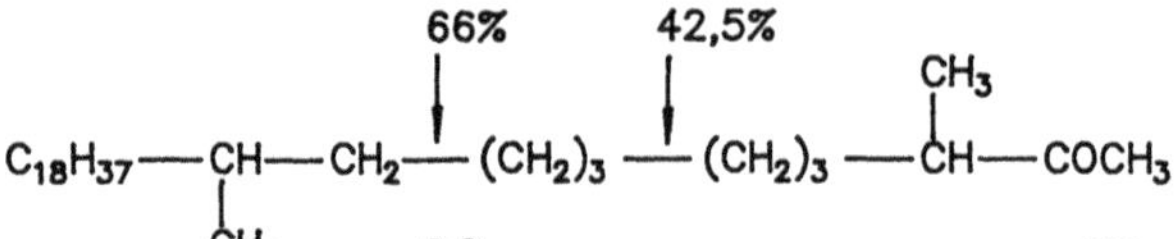

$$C_{18}H_{37}-\underset{CH_3}{\overset{}{CH}}-CH_2 \xrightarrow{66\%} (CH_2)_3 \xrightarrow{42,5\%} (CH_2)_3 -\underset{}{\overset{CH_3}{CH}}-COCH_3$$

26

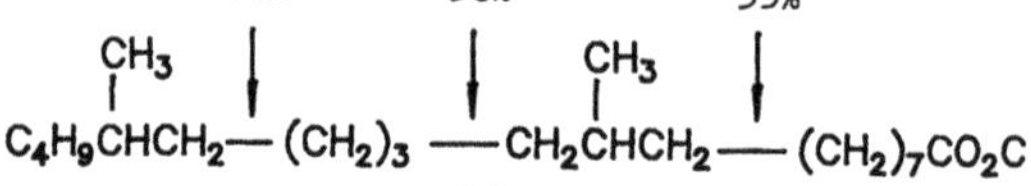

$$CH_3(CH_2)_6 \xrightarrow{46\%} CH_2\underset{CH_3}{\overset{}{CH}} \xrightarrow{31-38\%} (CH_2)_7CO_2CH_3$$

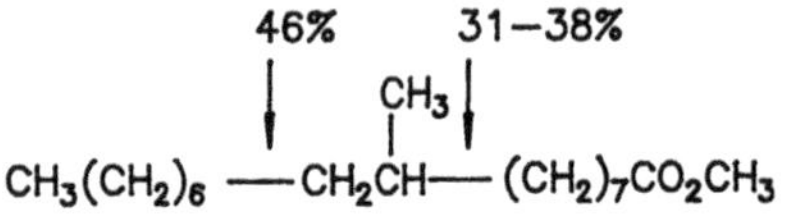

27

$$C_4H_9\underset{CH_3}{\overset{}{CH}}CH_2 \xrightarrow{36\%} (CH_2)_3 \xrightarrow{50\%} CH_2\underset{CH_3}{\overset{}{CH}}CH_2 \xrightarrow{55\%} (CH_2)_7CO_2C$$

28

Me-CH(Me)-CH$_2$-CH$_2$-C*H(Me)-(CH$_2$)$_3$-C*H(Me)-CH$_2$-Br

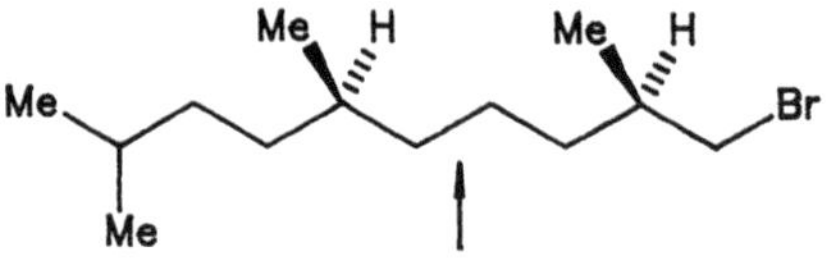

29

Muscalure **20**, the pheromone of the housefly has been prepared from oleic acid or erucic acid, similarly (Z)-11-heneicosene **21**, the synergist of muscalure was obtained [189]. The intermediate **22** for the pheromone of the Cabbage looper was prepared using (Z)-methyl-4-octenedioate [166b], that was obtained by partial ozonolysis of (Z,Z)-1,5-cyclooctadiene. Similarly disparlure **23**, the sex attractant of the gypsy moth, has been synthesized by two successive crossed-couplings with (Z)-4-octene dioate [191].

The optically active *Trogoderma* -pheromones (E)- and (Z)-**24** have been synthesized starting from (S)-citronellol and 4-pentynoic acid [192]. Alkatrienes **25**, sex attractants of Lepidoptera, were prepared by mixed Kolbe electrolysis with linolenoic acid [193]. **26**, the pheromone of the German cockroach *Blattela Germanica* has been prepared from 3-methylheneicosanoic acid [194]. (±) — Tuberculostearic acid (**27**) has been obtained in two successive electrolyses from methyl 2-methyl-succinate [195]. 10,16-Dimethyleicosanoic acid **28** was synthesized by three successive electrolyses [180]. The bromide **29**, a key intermediate in the synthesis of natural α-tocopherol has been prepared starting from (R)-(+)-citronellic acid [196]. The alkane **30** with three quaternary carbon atoms has been obtained by two successive cross-couplings [122]. In an alternative approach looplure **31**, the pheromone of the Cabbage Looper was made available by Kolbe electrolysis of (Z)-4-nonenoic acid with methyl glutarate [197]. Similarly the pheromone of the fruit pest insect *Dacus cucurbitae* **32** [198] and of the false codling moth **33** [199] have been prepared. Cross-coupling of (E)-3-hexenoic acid and levulinic acid provided a shorter route to brevicomin **34** [200], the sex attractant of the western pine beetle. Pure (E)- or (Z)-configurated unsaturated pheromones can be prepared by cross-coupling with 5-alkynoic acids and subsequent selective hydrogenation (Table 7) [167]. In the cross-coupling of perfluorinated acids with unsubstituted acids an excess of the weaker acid and a solvent with alkaline properties, e.g. pyridine, seems to be profitable [201]. The attempted cross coupling between trifluoroacetate and propionate does not lead to a mixed dimer but to the addition product of the CF_3-radical to ethylene formed from propionate [43b]. Useful intermediates for the synthesis of dicarba analogues of cystine peptides **35** have been prepared by mixed Kolbe-electrolysis of protected L- or D-glutamic acids [202]. Chiral building blocks **36–38** for synthesis have been obtained by cross-coupling with enantiomerically pure β-hydroxybutyric acid derivatives [203]. Chiral γ-lactones **39** have been prepared from (R)-3-cyclohexene-1-carboxylic acid [204].

By coelectrolysis of polymethacrylic acid with ε-acetaminocaproic acid or cyanoacetic acid the alkylacetamido- or cyanomethyl group can be grafted on to the main chain of the polymer [205].

6 Addition of Kolbe Radicals to Double Bonds

Kolbe radicals can be added to olefins that are present in the electrolyte. The primary adduct, a new radical, can further react by coupling with the Kolbe radical to an additive monomer I (Eq. 9, path a), it can dimerize to an additive dimer II (path b), it can be further oxidized to a cation, that reacts with a nucleophile to III (path c), or it can disproportionate (path d).

Oligomerization and polymerization of the olefin by addition of the primary adduct to further monomers occurs only in a few cases (see below). The radical concentration in the reaction layer in front of the electrode is so high, that the termination step and not the propagation step of the polymerization predominates. On the other hand as a result of the high radical concentration the dimerization of the Kolbe radical can compete strongly with the addition. With reactive olefins, e.g. butadiene, isoprene or styrene, generally good yields of adduct are obtained (Table 8, Nos. 1–10). With unreactive olefins, however, like cyclohexene, isobutene (Table 8, No. 11) the yields of addition product are low and Kolbe dimers predominate. However, for ethylene good yields have also been claimed (No. 12). The yields can be increased, when nucleophilic radicals, as alkyl radicals, are added to electrophilic olefins, as α,β-unsaturated carbonyl compounds or nitriles (Nos. 17–22). The same seems to be valid for the reaction of electrophilic radicals (CF_3, MeO_2CCH_2) with nucleophilic olefins, as enols or enolacetates (Nos. 23–27). With the trifluoromethyl radical, obtained from trifluoroacetic acid, also with electrophilic olefins fair yields of adduct have been found.

To some degree the ratio of additive monomer to additive dimer can be influenced by the current density. High current densities favor the formation of additive monomers, low ones these of additive dimers (Table 8, Nos. 4, 5). This result can be rationalized according to Eq. 9: At high current densities, which corresponds to a high radical concentration in front of the electrode, the olefin can trap only part of the Kolbe radicals formed. This leads to a preferred coupling to the Kolbe dimer and a combination of the Kolbe radical with the primary adduct to the additive monomer. At low current densities the majority of the Kolbe radicals are scavenged by the olefin, which leads to a preferential formation of the additive dimer.

In the addition to 1,3-dienes, e.g. butadiene, an intermediate allyl radical is formed, that couples to 1,1'-, 1,3'- and 3,3'-dimers (Eq. 10). The reactivity of the intermediate allyl radical is about 2.4 to 2.7 higher in the C1- than in the C3-position leading to a preferred formation of the 1,1'-dimer [206].

The ratio of C1- to C3-coupling is not influenced by the solvent but to some extent by the electrode material [227].

Additions of Kolbe radicals to dienes are reported in refs. [45, 215, 228] and in the reviews named in chap. 1.

Table 8. Addition of Kolbe-radicals to olefins

No.	Carboxylic acid	Olefin	Additive monomer (%-yield)	Additive dimer (%-yield)	Ref.
1	$CH_3O_2CCH_2CO_2H$	styrene	4	38[a]	206
2	$CH_3O_2CCH_2CO_2H$	butadiene	17	18	206
3	$CH_3O_2CCH_2CO_2H$	isoprene	1	43	206
4	EtO_2CCO_2H	butadiene ($i = 0.025$ A/cm^2)	4	66	206, 207
5	EtO_2CCO_2H	butadiene ($i = 0.66$ A/cm^2)	15	8	206
6	EtO_2CCO_2H	isoprene	8	59	206
7	$CH_3O_2C(CH_2)_4CO_2H$	butadiene	48	48	208, 209
8a	HO_2CCO_2H	butadiene	—	20	210
8b	HO_2CCO_2H	![OMe/OMe furanone]	62	34	211
9	$CH_3O_2CCH_2CO_2H$	α-methyl-styrene	—	46[b]	138b
10	CH_3CO_2H	styrene	—	15	212
11	EtO_2CCO_2H	isobutene	9	—	206
12	CH_3CO_2H	$CH_2=CH_2$	~70	—	213
13	Cl_3CCO_2H	$CH_2=CH_2$	—	—	214
14	HO_2CCO_2H	$CH_2=CH_2$	25	46	213
15	$CH_3O_2C(CH_2)_4CO_2H$	$CH_2=CH_2$	15	—	213
16	CH_3CO_2H	$Me_3CCH=CH_2$	46	—	215
17	CH_3CO_2H	$CH_2=C(CH_3)CHO$	—	80	216
18	CH_3CO_2H	$CH_2=CH-CO_2Et$	—	70	216
19	CH_3CO_2H	⌒=⌒–CO$_2$CH$_3$ / CO$_2$CH$_3$	21-high	—	217, 218
20	$PhCH_2CO_2H$	⌒=⌒–CO$_2$CH$_3$ / CO$_2$CH$_3$	24	—	219

#	Acid	Substrate	Yield	Ref.
21	CH_3CO_2H	![maleimide N-R, R: Et, Ph]	80–88	219, 218
22	RCO_2H R: CH_3, C_2H_5, i-C_4H_9	$Ph-SO_2-CH=CH_2$	37–69	220
23	CF_3CO_2H	$CH_3C(OAc)=CH-CO_2C_8H_{17}$	H_3C-C(O)-CH(CF_3)-C(O)-OC_8H_{17}	221
24	$CH_3O_2CCH_2CO_2H$	$CH_2=CHOEt$	64	222
25	CF_3CO_2H	1-pentene	— 35	223
26	CF_3CO_2H	$CH_2=C(CH_3)OAc$	18[a] 40	224
27	CF_3CO_2H	sulfolene (SO_2 ring)	31	225
28	CF_3CO_2H	$CH_2=CHCO_2CH_3$	5 50	223
29	CF_3CO_2H	$CH_2=CHCO_2H$	17 28	170
30	CF_3CO_2H	$CH_2=CHCF_3$	40	213
31	CF_3CO_2H	$CHF=CF_2$	30	213
32	CF_3CO_2H	$CH(CO_2R)=CH(CO_2R)$, R = Me, Et	30 10 41 27	213 225
33	CF_3CO_2H	$(CH_2)_n(CH=CHCO_2Et)_2$, n = 0–3	4–42	226

[a] Additional 16% disproportionation product, [b] 6% Product-type IV, Eq. 9

The electrolysis of methyl adipate in the presence of butadiene has received considerable attention, because it makes long chain diacids easily accessible. A total yield of 96% diester has been claimed for this reaction (Table 8, No. 7).

The addition of various Kolbe radicals generated from acetic acid, monochloroacetic acid, trichloroacetic acid, oxalic acid, methyl adipate and methyl glutarate to acceptors such as ethylene, propylene, fluoroolefins and dimethyl maleate is reported in ref. [213]. Also the influence of reaction conditions (current density, olefin-type, olefin concentration) on the product yield and product ratios is individually discussed therein. The mechanism of the addition to ethylene is deduced from the results of adsorption and rotating ring disc studies. The findings demonstrate that the Kolbe radicals react in the surface layer with adsorbed ethylene [229]. In the oxidation of acetate in the presence of 1-octene at platinum and graphite anodes, products that originate from intermediate radicals and cations are observed [230].

In some cases the polymerization of reactive olefins can be initiated by electrolysis of carboxylic acids. Monomers that have been polymerized this way are styrene [212],

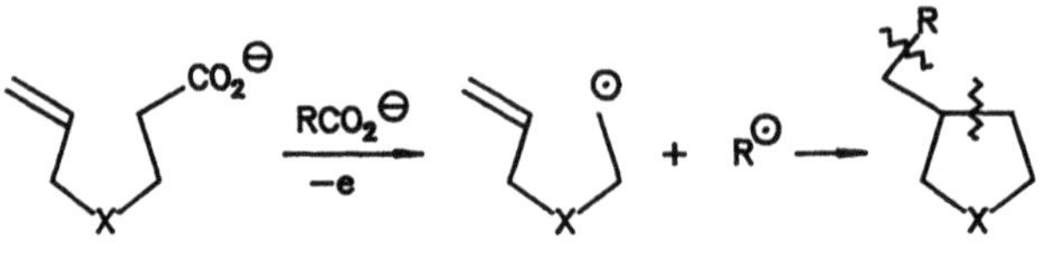

$$\text{(11)}$$

X= O, NCOR, C(CH$_3$)$_2$

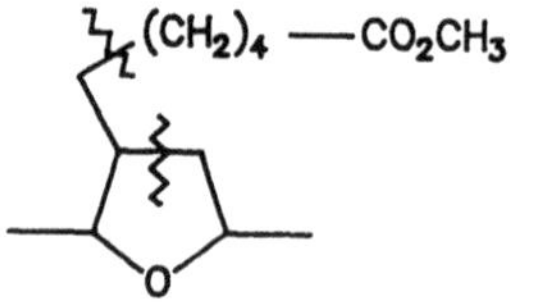

40: (CH$_2$)$_4$—CO$_2$CH$_3$ (O-containing ring)

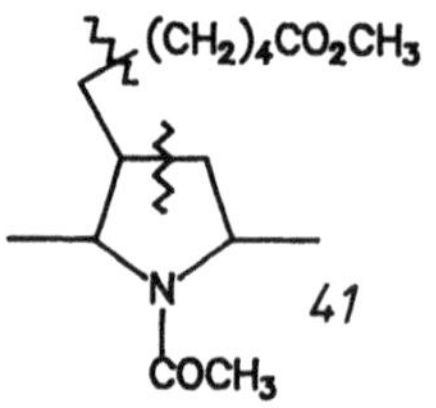

41: (CH$_2$)$_4$CO$_2$CH$_3$ (N-COCH$_3$ ring)

42: (CH$_2$)$_4$CO$_2$CH$_3$ (C(CH$_3$)$_2$ ring)

43: (CH$_2$)$_4$CO$_2$CH$_3$, NC (C(CH$_3$)$_2$ ring)

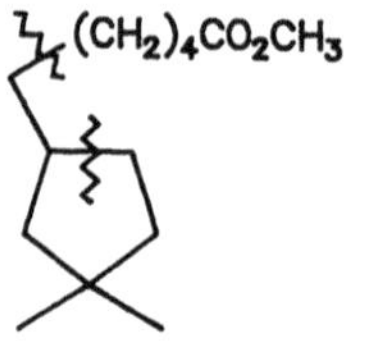

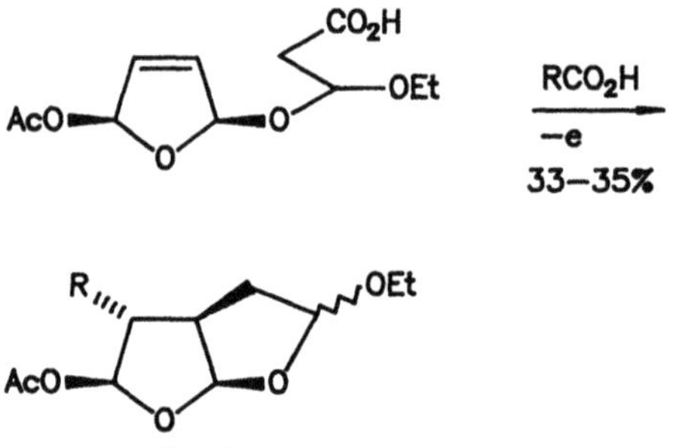

$$\text{(12)}$$

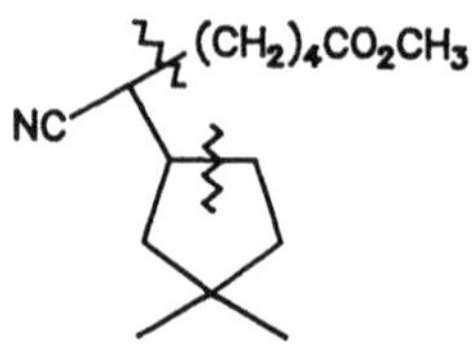

33–35%

R: CH$_3$, (CH$_2$)$_2$CO$_2$CH$_3$

acrylonitrile [212], vinyl acetate [231], methyl acrylate [231], vinylchloride [232], acrylic acid [232] and acrylamide [233].

Kolbe radicals can also be trapped by oxygen to yield dialkylperoxides, aldehydes, and ketones [97]. Furthermore methyl and trifluoromethyl radicals from acetic acid and trifluoroacetic acid are trapped, although inefficiently, by pyridine (3–20%) [234], benzotrifluoride and benzonitrile[235].

Alkyl radicals that add only in low yield to nucleophilic alkenes can be more efficiently trapped, if they react intramolecularly. Kolbe electrolysis of $\Delta^{6,7}$- and $\Delta^{7,8}$- unsaturated carboxylates leads to five membered and six membered rings in a 5- or 6-*exo*-trig cyclization [139]. Such an intramolecular cyclization was first reported by Weedon [138a]. In a systematic study on the Z/E-isomerization of unsaturated carboxylates with different distances of the double bond from the carboxylate group cyclopentanes and to a smaller extent cyclohexanes were formed by cyclization of the intermediate 5-hexenyl and 6-heptenyl radicals [140]. This reaction has been utilized for the efficient construction of heterocycles (Eq. 11, **40** (41%) [236], **41** (53%) [237], and carbocycles **42** (74%), **43** (75%) [238]. The reaction has been applied to the synthesis of a prostaglandine precursor (Eq. 12) [239]. The stereospecificity of this reaction can be rationalized by the *cis*-annulation of five-membered rings and the predominant coupling of the cyclized radical from the sterically unshielded site. Compared to chemical radical cyclizations [240] these electrochemical ring closures have the advantages that they join two carbon bonds in one step, whilst in the chemical reactions in most of the cases only one carbon-carbon and one carbon-hydrogen bond are formed. Secondly electrolysis avoids the toxic tributylin hydrides, that are mostly used as initiators in the chemical radical cyclizations.

7 Non-Kolbe Electrolysis to Carbenium Ions as Intermediates

The electrolysis of carboxylic acids in aqueous solution can lead to alcohols and esters as major reaction products [5, 241]. When electrolyses are performed in methanol or acetic acid methyl ethers or acetates can be found as side or major products. These observations led Walling and others [242] to suggest that in these cases the inter-

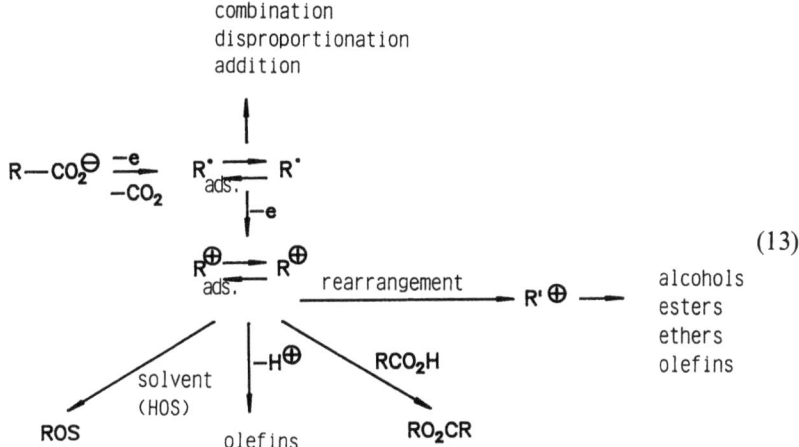

(13)

mediate Kolbe radicals are further oxidized to carbenium ions that undergo solvolysis and elimination. The oxidation of carboxylates to carbenium ions has been applied in mechanistic investigations and used in synthesis, this work is summarized in refs. [9, 18, 19, 25].

In order to be more precise the oxidation of carboxylates to radicals shown in Eq. 2, has to be extended by some additional steps outlined in Eq. 13. The free or adsorbed radical is further oxidized to a free or adsorbed carbenium ion. This can solvolyze to ethers, alcohols or esters, can lose a proton to form an olefin, can combine with the unreacted acid to an ester, or can undergo rearrangement or fragmentation prior to these reactions. This part of the anodic decarboxylation, leading to carbenium ions, is often called the Hofer-Moest reaction, when water is used as solvent, or more generally non-Kolbe electrolysis. The two electron oxidation can be recognized by a major portion of olefinic products, the formation of alcohols, esters, ethers, amides and especially by means of products that originate from rearrangement or fragmentation.

Low current densities suppress Kolbe-dimerization and promote the formation of alcohols or ethers [9, 31]. Non-Kolbe electrolysis is further favored by graphite as the electrode material [243, 244, 245, 246 b]. However, if nonporous carbon anodes (baked carbon, vitreous carbon, hard carbon) are used Kolbe dimers are formed in moderate to good yields [48]. This suggests that the different real surface areas of the carbon anodes lead to different current densities and thus different radical concentrations at the electrode. High concentrations at dense graphite favor the bimolecular dimerization, whilst low ones at soft graphite support the second electron transfer. With porous graphite even primary alkanoates can be oxidized to primary carbenium ions [52].

A mixture of water/pyridine appears to be the solvent of choice to aid carbenium ion formation [246]. In the Hofer-Moest reaction the formation of alcohols is optimized by adding alkali bicarbonates, sulfates [39] or perchlorates. In methanol solution the presence of a small amount of sodium perchlorate shifts the decarboxylation totally to the carbenium ion pathway [31]. The structure of the carboxylate can also support non-Kolbe electrolysis. By comparing the products of the electrolysis of different carboxylates with the ionization potentials of the corresponding radicals one can draw the conclusion that alkyl radicals with gas phase ionization potentials smaller than 8 eV should be oxidized to carbenium ions [8c] in the course of Kolbe electrolysis. This gives some indication in which cases preferential carbenium ion formation or radical dimerization is to be expected. Thus α-alkyl, cycloalkyl [247, 248], α-chloro [249] or bromo, α-amino, alkoxy, hydroxy, acyloxy, α,α-diphenyl more or less promote the oxidation of the radical to the carbenium ion. Besides electronic effects the oxidation seems also to be controlled by steric factors. In the oxidation of nuclear substituted phenylacetic acids even strongly electron withdrawing groups, such as CF_3, do not inhibit the oxidation of the radical. This has been interpreted by a lower extent of adsorption, from which the dimerization is assumed, due to steric shielding [31]. In summary, the experimental factors that favor the carbenium ion pathway are: low current density, a porous graphite anode, addition of foreign anions (e.g. perchlorate), a high pH and structural factors in the carboxylic acid as electron donating groups in the α-position and bulky substituents that retard adsorption of the radicals.

It is generally assumed that the carbenium ion is formed by oxidation of the intermediate radical (Eq. 13). There are, however, small differences in the product ratios of configurational isomers in the non-Kolbe electrolysis of **44/45** and **46/47** that could be interpreted as a result of a synchronous electrocyclic ring opening of an acyloxy cation (RCO_2^+) [249]. The differences are, however, small and do not point to RCO_2^+ as being a significant intermediate. The carbenium ions formed are assumed to be "hot". The product distribution is similar to that obtained from carbenium ions generated chemically by deamination and deoxygenation [243, 250, 251]. Compared to the chemically formed carbocations, the pH in the electrolysis can be chosen rather freely. This way rapid rearrangements can be partially intercepted by electrolysis at high pH [252], e.g. of the anodically generated pinacolyl cation [253]. The degree and rate of rearrangement seems to be independent of the electrode material, whether it is platinum or carbon [254] appears to have no effect.

Non-Kolbe electrolysis may lead to a large product spectrum, especially when there are equilibrating cations of about equal energy involved. However, in cases where the further reaction path leads to a particularly stabilized carbocation and either elimination or solvolysis can be favored, then non-Kolbe electrolysis can become an effiyient synthetic method. This is demonstrated in the following chapters.

8 Non-Kolbe Electrolysis of Carboxylic Acids to Ethers, Esters, and Alcohols

Carboxylic acids with an electron donating substituent in the α-position decarboxylate in a two-electron oxidation to carbocations (see chap. 7). These can react with the solvent (alcohol, acetic acid, water) or the unreacted carboxylate to ethers, esters, or alcohols (Eq. 14). In some cases the carbon skeleton rearranges, which is a clear indication of the cationic pathway.

$$R^1-\underset{\underset{R^2}{|}}{CH}-CO_2^\ominus \xrightarrow[-CO_2]{-2e} R^1-\underset{\underset{R^2}{|}}{CH}^\oplus \xrightarrow{HOS} R^1-\underset{\underset{R^2}{|}}{CH}-OS \qquad (14)$$

$SO = CH_3O, CH_3CO_2, HO$

Table 9. Preparation of ethers, esters and alcohols by non-Kolbe electrolysis of carboxylates

No.	Carboxylate	Conditions	Product	Yield (%)	Ref.
1	Adamantane-1-carboxylic acid	MeOH or EtOH, Pt	1-methoxy-, or 1-ethoxy-adamantane	50	248, 255a,b
2	Homoadamantane-1-carboxylic acid	MeOH or EtOH, Pt	1-methoxy-, or 1-ethoxy homo-adamantane	—	255c
3	(norbornyl)-CO$_2$H exo- or endo-	MeOH, Pt	(norbornyl)-OMe	35–40	242c
4	C$_6$H$_5$-C(cyclobutyl)-CO$_2$H	MeOH, Pt	C$_6$H$_5$-C(cyclobutyl)-OMe	33	256
5	(cyclohexenyl)-CO$_2$H	MeOH	(cyclohexenyl)-OMe	100	96
6	Ph$_2$C=C(CO$_2$Et)(CH$_2$CO$_2$H)	EtOH	Ph$_2$C=C(CO$_2$Et)(CH$_2$OEt)	69	257
			Ph$_2$C(OEt)–C(CO$_2$Et)=CH$_2$	12	
7	α-Methoxyphenylacetic acid	MeOH	benzaldehyde dimethylacetal	62	258
8	α-Methoxydiphenylacetic acid	MeOH	benzophenone dimethylacetal	74	258
9	α-Ethoxy-4-nitrophenylacetic acid	MeOH	4-nitrobenzaldehyde methyl ethylacetal	50	259
10	Triphenylacetic acid	MeOH	triphenylmethyl methyl ether	60	260
11	Diphenylacetic acid	MeOH, Et$_3$N	methyl benzhydryl ether	80	261
12	R^{1}R^{2}R^{3}CCOCO$_2$H R^1=R^2=R^3=CH$_3$ R^1=R^2=R^3=C$_5$H$_{11}$	MeOH, Pt	R^{1}R^{2}R^{3}CCO$_2$CH$_3$	50–60	262

#	Substrate	Conditions	Product	Yield	Ref.
13	furan-CO$_2$H	MeOH, NH$_4$Br, Et$_3$N, C	furan-CH(OMe)$_2$ / CO$_2$Me	78	263
14	furan-CO$_2$H	MeOH, Et$_3$N, Et$_4$NClO$_4$, Pt	furan-CH(OMe)$_2$ / CO$_2$Me	84	264
15	Phenylacetic acid	Pyr.-MeOH, C	benzyl methyl ether	93	119
16	Ph$_2$C(SPh)CO$_2$H	MeOH	benzophenone dimethylacetal	64	265
17	PhSO$_2$CH$_2$CO$_2$H	MeOH	PhSO$_2$CH$_2$OMe	57	266
18	norbornene-CO$_2$H	MeOH	norbornyl-OMe	56	242c
19	norbornene-SO$_2$Ph/CO$_2$H	MeOH	MeO–norbornyl–SO$_2$Ph	49	266
20	tetrahydropyranyl-R, CO$_2$H; R = Allyl, CH$_2$—CH=CHMe, CH$_2$—CMe=CH$_2$	MeOH, K$_2$CO$_3$, Pt	tetrahydropyranyl-R, OMe; exo/endo: 2/1	67–90	267
21	CH$_3$(CH$_2$)$_n$CH−CO$_2$H, SPh; n = 2,7	MeOH, LiClO$_4$, Pt	CH$_3$(CH$_2$)$_n$CH(OMe)$_2$	72–98	268
22	R^{1}CONHCHR^{2}CO$_2$H; R^1: Ph, Me, PhCH$_2$; R^2: H, Me	R^{3}OH; R^3: Me, Et, iPr, CH$_3$CO	R^{1}CONHCHR^{2}OR^3	38–91	269

Table 9. continued

No.	Carboxylate	Conditions	Product	Yield (%)	Ref.
23	R^1C–CO_2H with CO_2Et, NHCOMe	R^2OH	R^1C–OR^2 with CO_2Et, NHCOMe	79–96	270
	R^1: H, Me, Et, ^{i}Prop, PhCH$_2$	R^2: Me, Et, ^{i}Prop, CH$_3$CO			
24	Ph, CO$_2$Et, CO$_2$H, N-COCH$_3$ pyrrolidine	MeOH	Ph, CO$_2$Et, OMe, N-COCH$_3$ pyrrolidine	98	271
25	p-NO$_2$-C$_6$H$_4$-CO-D,L-Val-OH	MeOH	p-NO$_2$C$_6$H$_4$CONH–CH(OMe)CHMe$_2$	35	272
26	ZNH–CH(Me)–CO–NH–CHR^1–CO$_2$H	MeOH	ZNH–CH(Me)–CO–NH–CHR^1–OMe		273
	Z: PhCH$_2$O$_2$C; R^1: H, ^{i}Prop				
27	HO$_2$C⋯ pyrrolidine with OR^1, N–R^2	MeOH	MeO⋯ pyrrolidine with OR^1, N–R^2		274a
	R^1: SiMe$_2$^{t}But, R^2: CO$_2$Me				
	OH, HO$_2$C⋯ pyrrolidine, N–COCH$_3$	MeOH, Pt	OH, CH$_3$O⋯ pyrrolidine, N–COCH$_3$	97	274b

#	Starting material	Conditions	Product	Yield	Ref.
29	RO-[OMe, NHAc] CO$_2$H	AcOH, Pt	RO-[OMe, NHAc] OAc	89	275
30	H$_{13}$C$_6$-[furan]-CO$_2$H	AcOH/NaOAc	AcO, H$_{13}$C$_6$-[furanone]	57	276
31	R^1–C(NHCOR^2)(CO$_2$Et)–CO$_2$H; R^1: Me, Allyl, H, Et, Benzyl; R^2: Me, OBenzyl	AcOH/NaOAc	Me–C(NHAc)(CO$_2$Et)–OAc	40–91	277
32	CH$_3$(CH$_2$)$_n$–CHCl–CHCO$_2$H; n: 2,3	MeOH	CH$_3$(CH$_2$)$_n$CHCl–C(=O)Cl / CH$_3$(CH$_2$)$_n$–HCO	41–60	249
33	ROCH$_2$CO$_2$H; R: Me, Et	MeCN	ROCH$_2$COCH$_2$OR	47–66	278
34	R-C$_6$H$_4$-CH$_2$CO$_2$H; R: 4-MeO, 4-Cl, 3-MeO, H	DMSO	R-C$_6$H$_4$-CHO	41–78	279
35	HO$_2$C(CHOH)$_2$CO$_2$H	H$_2$O, W-anode	OHC–CHO	—	280

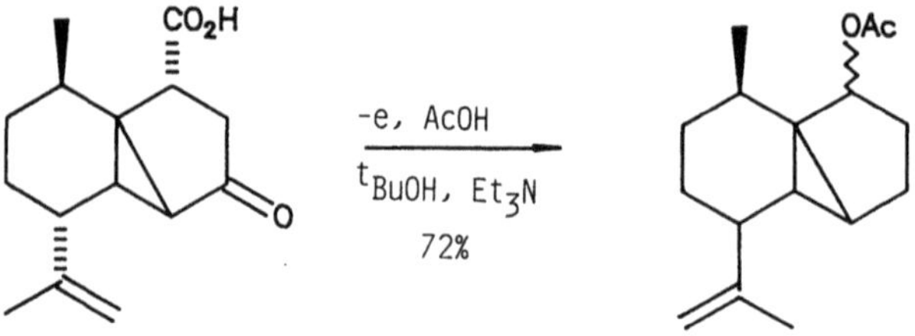

[Structural scheme for Eq. (15): tricyclic carboxylic acid → acetate, -e, AcOH, tBuOH, Et₃N, 72%]

(15)

[Structural scheme for Eq. (16): cyclic alkenyl carboxylic acid (n: 4–13) → allylic acetate, Pt, -e, -CO₂, CH₃CO₂H, tBuOH, Et₃N, ~80%]

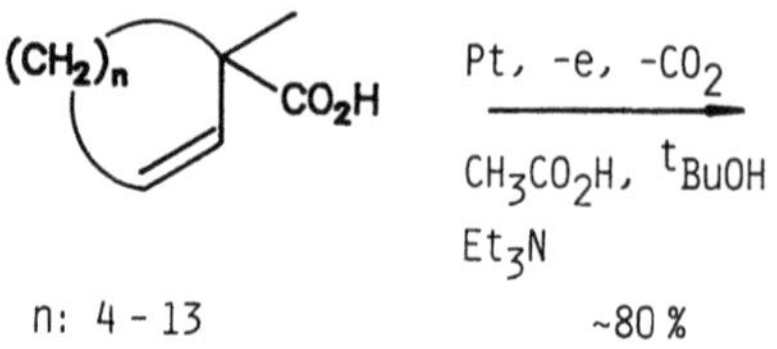

(16)

In this way bridgehead carboxylates have been converted into the corresponding ethers (Table 9, Nos. 1–2). Both *exo-* and *endo-*2-norbornylcarboxylic acid yield the *exo-*2-methyl ether indicating that a bridged norbornyl cation is the intermediate (Table 9, No. 3). Steric reasons can prevent the rearrangement as in the conversion of the tricyclic carboxylic acid to the corresponding acetate (Eq. 15) [281]. In other cases the intermediate cyclopropylcarbinyl cation undergoes rearrangement (see also chap. 11). Carboxylic acids formed by a Stobbe condensation can react to an allyl cation that solvolyses in the 1- and 3-position (Table 9, No. 6). In some cases the allyl cation undergoes a regioselective acetolysis (Eq. 16) [282]. In the norbornenyl system the intermediate cation can add intramolecularly to the double bond leading to nortricyclene derivatives (Table 9, Nos. 18, 19).

α-Phenyl substituted carboxylic acids react selectively, because the intermediate phenylmethyl cations cannot undergo elimination or rearrangement (Table 9, Nos. 4, 7–11, 15). α-Ketoesters can be converted to esters possibly via intermediate acylium cations (Table 9, No. 12). The α-alkoxy group is very effective in directing the decarboxylation towards the cation, whose solvolysis leads to acetals (Table 9, Nos. 13, 14, 20, 28, 29, 32). Cyclic ketals of β-ketocarboxylic acids react to 2-methoxy-1,4-dioxenes (Eq. 17) [283]. Uronic acid has been converted to a pentaacetate (Eq. 18) [275].

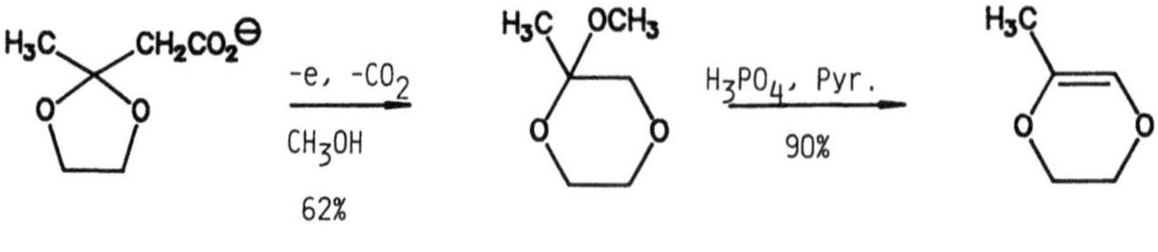

[Structural scheme for Eq. (17): dioxolane carboxylate → 2-methoxy-1,4-dioxene, -e, -CO₂, CH₃OH, 62%; then H₃PO₄, Pyr., 90%]

(17)

[Structural scheme for Eq. (18): uronic acid → pentaacetate, 1. CH₃OH, Et₂NH, C; 2. NaBH₄; 3. Ac₂O, 55%]

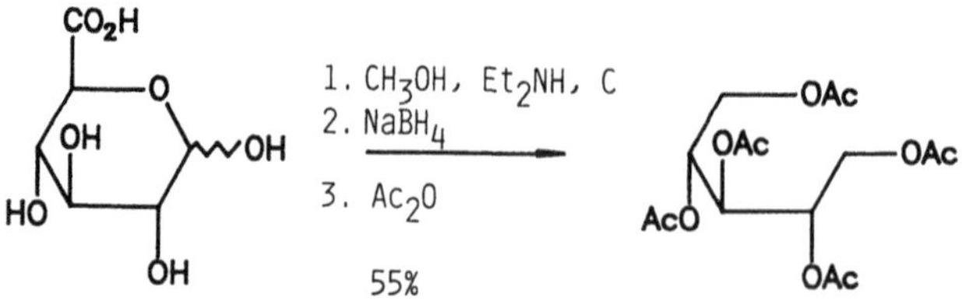

(18)

[Eq. (19): scheme with R: S-Benzthiazol: 61%; : CO_2H : 55%; reagents Et_3N, C, HOAc, ^tBuOH]

[Eq. (20): scheme with $-2e, CO_2$, CH_3OH, 58–71%]

R = C_2H_5, $CH_2CH=CH_2$, $CH_2C_6H_5$

2-Furane carboxylic acids are first oxidized to the 1,4-dimethoxy compound and then decarboxylated (Table 9, Nos. 13, 14, 30). The current density and thus the potential at the cathode can determine whether (Table 9, No. 13) or not (Table 9, No. 14) a subsequent reductive hydrogenation at the cathode occurs. The thio- and even the sulfonyl group can promote the further oxidation to the cation (Table 9, Nos. 16, 17, 21) leading to acetals in these cases. When this reaction is combined with an α-alkylation the α-thio carboxylic acid can be used as a d_1-synthon (Eq. 19) [284a–c]. The malonate unit can be employed in the same way (Eq. 19) [284d]. (S)-Malic-acid derivatives are transformed into enantiomerically pure alkylmalonaldehyde esters (Eq. 20) [285]. Many examples are reported for the decarboxylation of α-amino acids to the corresponding amino acetals (Table 9, Nos. 22–27, 31).

[Eq. (21): scheme with $-e$, Pt, CH_3OH; then allyl-TMS, $TiCl_4$, 75%, >97% ds; R = $Si(CH_3)_2{}^tBu$]

n: 1,2; R^1: H, CH_3, Ph, OEt; R^2: CH_3, C_2H_5, H, $^iC_3H_7$

4-Hydroxy-L-prolin is converted into a 2-methoxypyrrolidine. This can be used as a valuable chiral building block to prepare optically active 2-substituted pyrrolidines (2-allyl, 2-cyano, 2-phosphono) with different nucleophiles and employing $TiCl_4$ as Lewis acid (Eq. 21) [286]. Using these latent N-acylimmonium cations (Eq. 22) [287] (Table 9, No. 31), 2-(pyrimidin-1-yl)-2-amino acids [288], and 5-fluorouracil derivatives [289] have been prepared. For the synthesis of β-lactams a 4-acetoxyazetidinone, prepared by non-Kolbe electrolysis of the corresponding 4-carboxy derivative (Eq. 23) [290], proved to be a valuable intermediate. O-Benzoylated α-hydroxyacetic acids are decarboxylated in methanol to mixed acylals [291]. By reaction of the intermediate cation, with the carboxylic acid used as precursor, esters are obtained in acetonitrile (Eq. 24) [292] and surprisingly also in methanol as solvent (Table 9, No. 32). Hydroxy compounds are formed by decarboxylation in water or in dimethyl sulfoxide (Table 9, Nos. 34, 35).

The various carbenium ions: *tert*-alkyl, bridgehead-, norbornyl-, allyl-, benzyl- or cyclopropylcarbinyl-cations, which are assumed to be intermediates in these decarboxylations are compiled in ref. [293].

9 Non-Kolbe Electrolysis of Carboxylic Acids to Acetamides

Non-Kolbe electrolysis of carboxylic acids in acetonitrile/water leads to acetamides as main products [294] (Table 10). The mechanism has been investigated by using ^{14}C-labeled carboxylic acids. The results are rationalized by assuming a reaction layer rich of carboxylate resulting in the formation of a diacylamide that is hydrolyzed

(Eq. 25) [295]. A similar mechanism has also been proposed for the electrolysis of isobutyric and pivalic acid in acetonitrile [296]. As the intermediate alkyl cation can rearrange and the intermediate iminium cation can furthermore react with the starting carboxylic acid three different amides can be isolated (Eq. 26) [295a]. The portion of the diacylamide can be considerably increased if the electrolyte consists of acetonitrile/acetic acid [295b].

$$Me_3CCO_2^{\ominus} \xrightarrow[-CO_2]{-2e} Me_3C^{\oplus} \xrightarrow{MeCN} Me_3C-N\equiv CMe^{\oplus}$$

$$\xrightarrow{^{14}CH_3-CO_2^{\ominus}} Me_3C-N=C-Me \longrightarrow Me_3C-NCOMe \quad (25)$$
$$\underset{\underset{CH_3}{|}}{\overset{^{14}O}{\underset{\diagdown}{C}}\diagup O} \qquad\qquad\qquad \underset{^{14}COMe}{|}$$

$$\xrightarrow{H_2O} Me_3CNHCOCH_3^{14} + Me_3CNHCOCH_3$$

$$Me_2CH-CH_2CO_2^{\ominus} \xrightarrow[MeCN,H_2O]{-2e,Pt} Me_3C^{\oplus} + MeCH^{\oplus}-Et \quad (26)$$

MeCONHCMe$_3$	Me$_2$CH—CH$_2$C(O)NHCMe$_3$	MeCON(COCH$_2$CHMe$_2$)CH(Me)Et
10%	37%	53%

At a graphite anode and with potassium valerate analogous products were obtained, the yields were at a maximum with a water content between 10% and 30%. When the percentage of water was increased, larger amounts of butanols were formed. With increasing chain length of the carboxylic acid $CH_3(CH_2)_nCO_2H$ (n: 6, 10) the amount of amide found decreased and the Kolbe dimer became the major product. This has been attributed to a higher concentration of carboxylates at the electrode surface due to a stacking effect, whereby radical coupling is favoured [52].

The rearrangement of the intermediate alkyl cation by hydrogen or methyl shift and the cyclization to a cyclopropane by a CH-insertion has been studied by deuterium labelling [298]. The electrolysis of cyclopropylacetic acid, allylacetic acid or cyclobutanecarboxylic acid leads to mixtures of cyclopropylcarbinyl-, cyclobutyl- and butenylacetamides [299]. The results are interpreted in terms of a rapid isomerization of the carbocation as long as it is adsorbed at the electrode, whilst isomerization is *inhibited by desorption*, which is followed by fast solvolysis.

Table 10. Acetamides by non-Kolbe electrolysis of carboxylates in acetonitrile

No.	Carboxylic acid	Product	Yield (%)	Ref.
1	Me_3CCO_2H	$Me_3CNHCOCH_3$	40, 80	294a, 295
2	$Me_2\overset{\mid}{C}CO_2Me$ $Me_2\overset{\mid}{C}CO_2H$	$Me_2\overset{\mid}{C}-CO_2Me$ $Me_2\overset{\mid}{C}-NHCOMe$	25	294a
3	cyclohexyl-CH$_3$/CO$_2$H	cyclohexyl-CH$_3$/NHCOCH$_3$	68	295
4	$C_2H_5C(CH_3)_2CO_2H$	$C_2H_5C(CH_3)_2NHCOCH_3$	63	295
5	$(CH_3)_2CHCO_2H$	$(CH_3)_2CH-N-COCH_3$ $\qquad\qquad\overset{\mid}{C}OCH(CH_3)_2$	50	295, 296
6	cyclohexenyl-CO_2H	cyclohexenyl-NR^1R^2	54	
		$R^1 = H, R^2 =$ cyclohexenyl-CO	20	
		$R^1 = CH_3CO, R^2 =$ cyclohexenyl-CO	45	
		$R^1 = CH_3CO, R^2 = H$	20	
7	$R^1-\overset{O}{\overset{\|}{C}}-\overset{Me}{\underset{R^2}{\overset{\mid}{C}}}-CO_2H$ R^1 = Me, i-Prop R^2 = Me, Et R^3 = Me, Et, i-Prop	$R^1-\overset{O}{\overset{\|}{C}}-\overset{Me}{\underset{R^2}{\overset{\mid}{C}}}-NHCOR^3$	18–50	104b
8	norbornenyl-CO_2H	norbornyl-NHCOMe	35	297

10 Conversion of Carboxylic Acids into Olefins by Non-Kolbe Electrolysis

Carboxylic acids can be converted into olefins, when there is a leaving group such as H (Eq. 27), $SiMe_3$, SPh or CO_2H in the β-position. The olefin is formed, when the carbocation, that is generated by decarboxylation, undergoes a subsequent E1-elimination. Some examples are summarized in Table 11 (Nos. 1–10).

$$\overset{\ominus}{O_2C}\diagdown C-\overset{H}{\underset{}{C}}\diagup \xrightarrow[-CO_2]{-2e} \diagdown \overset{\oplus}{C}-\overset{H}{C}\diagup \xrightarrow{-H^{\oplus}} \diagdown C=C\diagup \qquad (27)$$

Reaction No. 5 (Table 11) is part of a synthetically useful method for the alkylation of aromatic compounds. At first the aromatic carboxylic acid is reductively alkylated by way of a Birch reduction in the presence of alkyl halides, this is then followed by an eliminative decarboxylation. In reaction No. 9 decarboxylation occurs probably by oxidation at the nitrogen to the radical cation that undergoes decarboxylation (see pseudo-Kolbe electrolysis, chap. 14). Such a mechanism seems also to be involved in the decarboxylation/desulfenylation (Eq. 28) [318]. There sulfur appears to be oxidized to a radical cation that cleaves to the *exo*-methylene lactone, carbon dioxide a proton and a thiyl radical (reaction a); in the direct electrolysis the yield is much lower (reaction b). In the eliminative decarboxylations of cycloalkanecarboxylic acids (seven- to eleven-membered rings), besides the 1,2-elimination, a 1,3-elimination to cyclopropanes and a transannular 1,5-elimination occur to some extent [247, 319].

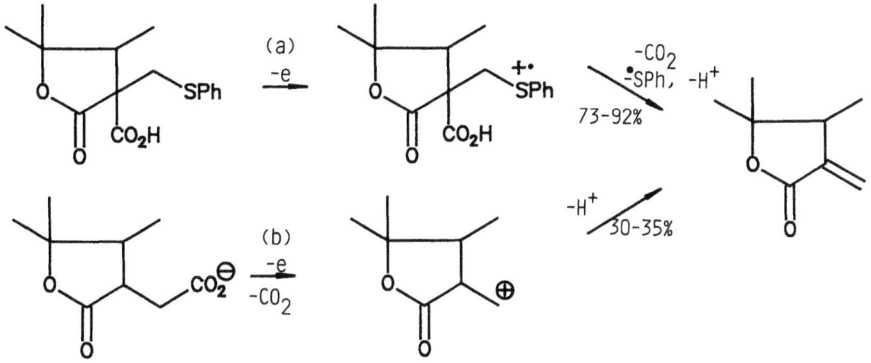

(28)

In β-trimethylsilylcarboxylic acids the non-Kolbe electrolysis is favored as the carbocation is stabilized by the β-effect of the silyl group. Attack of methanol at the silyl group subsequently leads in a regioselective elimination to the double bond (Eq. 29) [307, 308]. This reaction has been used for the construction of 1,4-cyclohexadienes. At first Diels-Alder adducts are prepared from dienes and β-trimethylsilylacrylic acid as acetylene-equivalent, this is then followed by decarboxylation-desilylation (Eq. 30) [308]. Some examples are summarized in Table 11, Nos. 12–13.

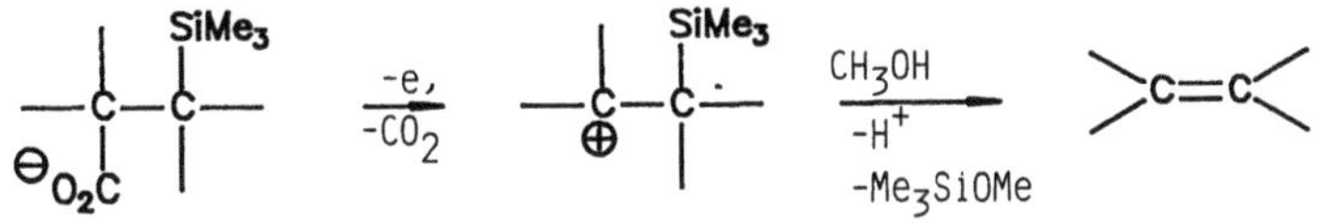

(29)

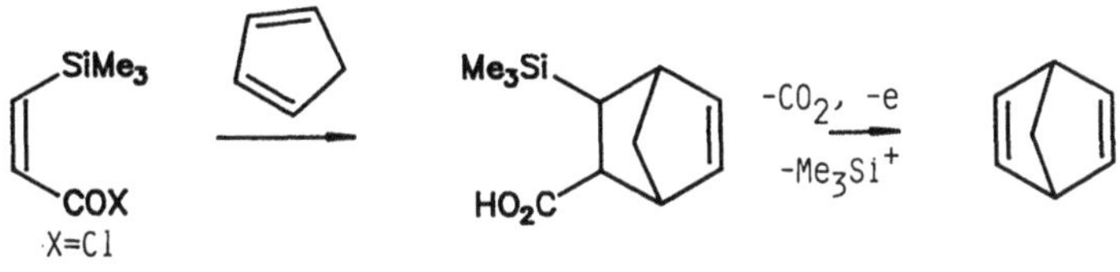

(30)

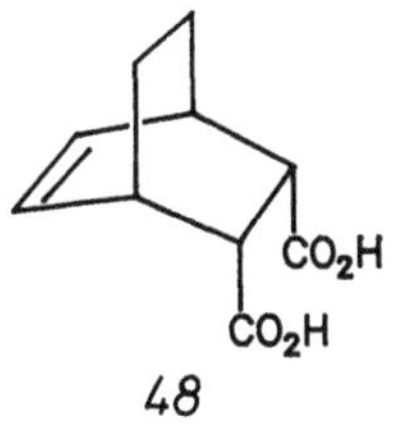

48 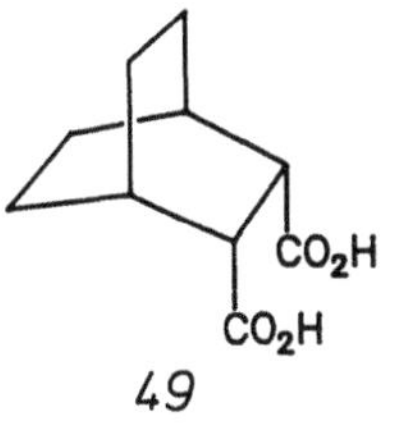   49

Bisdecarboxylation of vicinal dicarboxylates is the older method of converting vicinal diacids into olefins. The reaction can be combined with a [4 + 2]- or [2 + 2]-

127

Table 11. Conversion of Carboxylic acids into Olefins

No.	Carboxylic acid	Reaction Condition	Product	Yield (%)	Ref.
1	$MeCOC(CH_3)_2CO_2H$	H_2O, KOH	$MeCOC(CH_3)=CH_2$	90	104b, 303
2	(structure: 2-oxocyclopentane with $(CH_2)_4Br$ and CO_2H substituents)	H_2O KOH	(structure: 2-(4-bromobutyl)cyclopent-2-enone)	50	104b, 303
3	(steroidal structure with CO_2H and HO groups)	Pyr./H_2O, Et_3N, C	(steroidal butenolide with HO group)	57–63	300
4	(bicyclic lactone with CO_2H)	Pyr./H_2O, Et_3N, Pt	(bicyclic lactone olefin)	91	301
5	(cyclohexane with R^1, R^2, R^3, CO_2H); R^1 = iso-Prop, R^2 = Me, MeO, R^3 = H; R^1 = Me $(CH_2)_4$, R^2 = R^3 = OMe	MeOH, MeONa	(benzene with R^1, R^2, R^3)	60–70	302
6	(pyrazoline with H_5C_6, C_6H_5, CO_2H)	CH_3OH, Et_4NBF_4	(pyrazole with H_5C_6, C_6H_5)	86	304

	Substrate	Conditions	Product	Yield (%)	Ref.
7	R¹, R² on pyrrole with CO₂H, COCH₃ R¹ = C₆H₅, R² = CO₂Et R¹ = C₆H₅, R² = CO₂Et	H₂O, THF, KOH	R¹, R² on pyrrole with COCH₃	86–91	271
8	barbituric acid-CO₂H derivative	C, NaOMe, MeOH	uracil	91	305
9	tetrahydro-β-carboline-CO₂H, NH-CO₂H	MeOH, H₂O, KH₂PO₄, C	β-carboline methyl	75	306
10	PhSO₂CH₂CH(CO₂H)-	Pyr./MeOH (9:1), C	Ph–SO₂CH₂CH=CH₂ Ph–SO₂–CH=CH–CH₃ (1:1)	80	266
11	Me₃SiCH₂CHCO₂H | R R: C₁₂H₂₅, Br(CH₂)₆, etc.	MeCN, EtOH, C	RCH=CH₂	65–87	307
12a	4-methyl-2-(trimethylsilyl)cyclohexane-1-carboxylic acid	MeCN, EtOH, C	H₃C-C₆H₄ (toluene)	67	308
12b	norbornene-SiMe₃, CO₂H	MeCN, EtOH, C	norbornadiene	76	308

Table 11. continued

No.	Carboxylic acid	Reaction Condition	Product	Yield (%)	Ref.
12c		MeCN, EtOH, C		71	308
13		MeCN, EtOH, C		82	308
14		90% Pyr./H$_2$O, Pt		35	309, 310a
15		90% Pyr./H$_2$O, Pt		40–60	311
16		Pyr., MeCN, Et$_3$N		65	312
17		—		65	313
18		—		—	314

	Substrate	Conditions	Product	Yield %	Ref.
19	(diacid structure)	90% Pyr./H_2O	(diene structure)	35	310
20	(diacid structure)	90% Pyr./H_2O	(bicyclic structure)	48	309, 310
21	(diacid structure)	90% Pyr./H_2O	(bicyclic structure)	63	310a
22	(diacid structure)	90% Pyr./H_2O	(norbornene structure)	15	309
23	(diacid structure)	90% Pyr./H_2O	(benzo ketone structure)	51	315
24	(diacid structure)	90% Pyr./H_2O	(cyclohexenyl CO_2Me)	35	96
25	(triester diacid)	MeOH, MeONa, Pt	(tricyclic diester)	50	317
26	(tetraester diacid)	90% Pyr. H_2O, Pt	(bicyclic tetraester)	15	316

cycloaddition, e.g. with maleic anhydride, whereby unsaturated polycyclic hydrocarbons can be prepared in a few steps (Table 11, Nos. 14–25). 90% Pyridine:triethylamine:water as electrolyte, and platinum as electrode appear to favor this elimination. The optimum choice of conditions has been discussed in ref. [315]. The addition of the radical scavenger 4-*tert*-butyl-catechol to the electrolyte can improve the yield [312]. The anodic bisdecarboxylation is in most cases better than the corresponding oxidation with lead tetraacetate. When reaction No. 14 (Table 11) is conducted with lead tetraacetate only 3% of the product is yielded, sometimes, however, both methods are equivalent [320a]. For the dicarboxylic acids **48** and **49** the influence of the solvent, electrode material, current density, the kind of base and additives have been studied systematically [320b]. Out of the solvents acetonitrile, dimethylformamide, sulfolane, pyridine, 2,6-lutidine or pyridine/water (1:9) the latter leads to the highest yields and lowest passivation. From the anode materials platinum, gold, lead dioxide, graphite or glassy carbon the highest yields are obtained with platinum. At graphite, double bonds are apparently additionally oxidized. Higher current densities also support the reaction. The structure of the base (triethylamine or 1,4-diazabicyclo[2.2.2]octane) influences strongly the degree of passivation but not the yield. Addition of copper(II)acetate lowers the yield, that of lithium perchlorate inhibits the reaction.

As the mechanism, a radical and a cationic pathway are conceivable (Eq. 31). The stereochemical results with *rac*- or *meso*-1,2-diphenyl succinic acid, both yield only *trans*-stilbene [321], and the formation of a tricyclic lactone **51** in the decarboxylation of norbornene dicarboxylic acid **50** (Eq. 32) [309] support a cation (path b, Eq. 31) rather than a "biradical" as intermediate (path a).

$$(31)$$

$$(32)$$

However, the inhibition of the reaction by lithium perchlorate, that strongly favors the cationic pathway (see chap. 2, 7), contradicts this assumption. With regard to yield and the degree of passivation the decarboxylation/desilylation appears to be a better choice than the bisdecarboxylation for the construction of unsaturated polycyclic compounds (see for example Table 11, No. 12b and No. 22).

In the bisdecarboxylation of the cyclobutenedicarboxylic acid **52**, products are obtained whose formation possibly involves a cyclobutadiene intermediate [322]. A case of a 1,3-bisdecarboxylation has been reported in the preparation of a bicyclobutane (Table 11, No. 26). An elimination, that involves the cleavage of an carbon-oxygen bond after the decarboxylation, has been observed with the carboxylic acid **53** (Eq. 33) [282].

(33)

11 Rearrangement of Intermediate Carbocations in Non-Kolbe Electrolysis

The carbocations generated by non-Kolbe electrolysis can rearrange by alkyl, phenyl or oxygen migration. The migratory aptitudes of different alkyl groups have been studied in the rearrangement of α-hydroxy carboxylic acids (Eq. 34) [323].

$$R^1R^2\overset{OH}{\underset{|}{C}}-CH_2CO_2^{\ominus} \xrightarrow[-CO_2]{-2e} R^1R^2\overset{OH}{\underset{|}{C}}-CH_2^{\oplus} \xrightarrow{-H^{\oplus}} R^1\overset{O}{\underset{\|}{C}}-CH_2-R^2 \quad (34)$$

Migratory aptitude: vinyl > iso-propyl > cyclopropyl > benzyl

Rearrangements in the course of a decarboxylation can be taken as strong indicator for the involvement of carbocations. Anodic oxidation of *exo*- or *endo*- norbornane-2-carboxylic acid produced in both cases only the *exo*-norbornyl methyl ether (Table 12, No. 1) [242c]. *Exo*- or *endo*-5-norbornene-2-carboxylic acid react by double bond participation and subsequent rearrangement to 3-methoxynortricyclene (Table 12, No. 2). The kind of products was the same as obtained in solvolysis [329]. The product in reaction No. 3 (Table 12), has been used as intermediate for a methyl (±)-jasmonate synthesis. The three isomeric carboxylic acids in reaction No. 4 (Table 12) decarboxylate by rearrangement to the same product mixture. An analogous observation is made for the anodic oxidation of cyclopropylacetic acid, cyclobutylcarboxylic acid and allylacetic acid [252]. The anodic oxidation of cyclopropylcarboxylic acid can lead to Kolbe dimers (see chap. 4) and/or cationic products by ring opening to an allyl cation that undergoes methanolysis or acetolysis depending on the electrolyte [330].

In the electrolysis of quinuclidine-2-carboxylic acid decarboxylation occurs without rearrangement (Table 12, No. 5). Stereospecifically substituted cyclopentanes, that

Table 12. Rearrangements by non-Kolbe electrolysis

No.	Carboxylic acid	Reaction conditions	Product	Yield (%)	Ref.
1	norbornyl-CO₂H (exo or endo)	MeOH, Et₃N	norbornyl-OMe	35–40	242c
2	norbornenyl-CO₂H (exo or endo)	MeOH, Et₃N	nortricyclyl-OMe	56	242c
3	norbornyl-CO₂Me/CO₂H	AcOH, t-BuOH, Et₃N	MeO₂C–…–OAc	56	324
4	camphor-CO₂H	MeOH, MeONa	OMe product	74–82	325
4	bornyl-CO₂H		exo-methylene product	15–25	
5	N-azabicyclo-CO₂H	MeOH, MeONa	N-azabicyclyl-OMe	43	326

6	(structure with R¹, R², CO₂Me, CO₂H; R¹=R²=Me, H)	MeOH, MeONa	(structure with R², H, CO₂Me, R¹, HO, O)	87	327
7	(structure with CO₂H, O, O)	MeOH, MeONa	(structure with CO₂Me, OH, MeO, O)	49–58	328

Table 13. Ring Extension of alicyclic β-hydroxycarboxylic acids by non-Kolbe electrolysis

No.	Carboxylic acid	Products	Reaction conditions	Yield (%)	Ref.
1	$(CH_2)_n$ C(CH$_2$CO$_2$H)(OH)	$(CH_2)_{n+1}$ C=O	Pt, CH$_3$CN		
			n = 4:	54–63	242c
			n = 5:	54–63	242c
			C, Pyr./H$_2$O		
			n = 9:	82	335
			n = 11:	82	335
2	decalin-OH-CH$_2$CO$_2$H	decalin-fused cycloheptanone (two isomers)	C, DMF	56	336
3	decalin-OH-CH(Me)CO$_2$H	decalin-fused methylcycloheptanone (two isomers)	C, DMF	58	336
4	macrocycle with OH, CH$_2$CO$_2$H, CH$_3$	macrocyclic ketone with CH$_3$	MeOH, KOH	30	323
		macrocyclic ketone isomer	MeOH, KOH	5	

may serve as intermediates for the synthesis of cyclopentanoids, can be prepared by decarboxylation of oxa-bicycloheptanecarboxylic acids (Table 12, Nos. 6, 7). Migration of oxygen has been observed in the decarboxylation of glycidic acids [331, 332] and acetals of β-oxocarboxylic acids [333]. Rearrangement of the phenyl group was encountered in the decarboxylation of 3,3-diphenylpropionic acid, which afforded in acetic acid 1,2-diphenylethylacetate as the major product [334].

Non-Kolbe electrolysis of alicyclic β-hydroxy carboxylic acids offers interesting applications for the one-carbon ring extension of cyclic ketones (Eq. 35) [242c]. The starting compounds are easily available by Reformatsky reaction with cyclic ketones. Some examples are summarized in Table 13. Dimethylformamide as solvent and graphite as anode material appear to be optimal for this reaction.

$$(CH_2)_n \hspace{-0.2cm}\diagup\hspace{-0.2cm}C=O \quad \xrightarrow[\substack{1.\ BrZnCH_2CO_2Me \\ 2.\ OH^- \\ 3.\ H^+}]{} \quad (CH_2)_n \hspace{-0.2cm}\diagup\hspace{-0.2cm}C \diagup \substack{CH_2CO_2H \\ OH} \tag{35}$$

$$\xrightarrow[-2e,\ -CO_2]{C,\ DMF} \quad (CH_2)_n \hspace{-0.2cm}\diagup\hspace{-0.2cm}C \diagup \substack{CH_2^\oplus \\ OH} \quad \xrightarrow{-H^+} \quad (CH_2)_{n+1} \hspace{-0.2cm}\diagup\hspace{-0.2cm}C=O$$

With unsymmetrical cyclic ketones, however, mixtures due to similar migratory aptitudes of the different groups are obtained (Table 13, Nos. 2, 3). The rearrangement has also been used as key step in a d,l-muscone synthesis (Table 13, No. 4).

12 Fragmentation of Carboxylic Acids by Non-Kolbe Electrolysis

Non-Kolbe electrolysis of carboxylic acids can be directed towards a selective fragmentation, when the initially formed carbocation is better stabilized in the γ-position by a hydroxy or trimethylsilyl group. In this way the reaction can be used for a three-carbon (Eq. 36) [335] (Table 14, No. 1) or four-carbon ring extension (Eq. 37) [27] (Table 14, Nos. 2–4). Furthermore it can be employed for the stereo-

(36)

specific construction of cis- or trans-disubstituted cyclopentenes from 6-hydroxy-norbornane-2-carboxylic acids (Eq. 38) [338] (Table 14, Nos. 5, 6). Similarly it has been used for a stereospecific synthesis of a tetraalkylcyclohexane (Table 14, No. 7). The anodic fragmentation of α-acyloxycarboxylic acids: $RCO_2CR^1R^2CO_2H$ (R = Me, EtO; R^1, R^2 = H, Me, Ph) yields ketones R^1COR^2 and products derived from acyl cations: RCO^+ [341].

13 The Pseudo-Kolbe Reaction

Pseudo-Kolbe electrolysis is the name given to anodic decarboxylations where the electron transfer does not occur from the carboxylate but from a group attached to it [31]. These oxidations are characterized by potentials that are much lower than the critical potential for the Kolbe electrolysis. The salt of p-methoxyphenylacetic acid can be oxidized in methanol to afford the corresponding methyl ether as the sole product. The low oxidation potential of 1.4 V (sce) suggests, that the electron is being transferred from the aromatic nucleus (Eq. 39) [31].

Table 14. Fragmentation of carboxylic acids by non-Kolbe electrolysis

No.	Carboxylic acid	Reaction conditions	Product	Yield (%)	Ref.
1		C-anode, MeCN/EtOH, 100% neutralization	a)	n = 3 6 n = 4 47 n = 10 8	335
			a)	n = 3 39 n = 4 15 n = 10 52	
2		MeOH		44 35	337 335
3		CH_3CN/EtOH		85	27
4		CH_3CN/EtOH		70	27
5		C-anode, MeCN/EtOH		38	338
6		MeOH, C-anode		63	338
7		MeOH, Pt		34–38	339
8		MeOH, NeONa		53	340
				17	

[a] After hydrogenation of the product

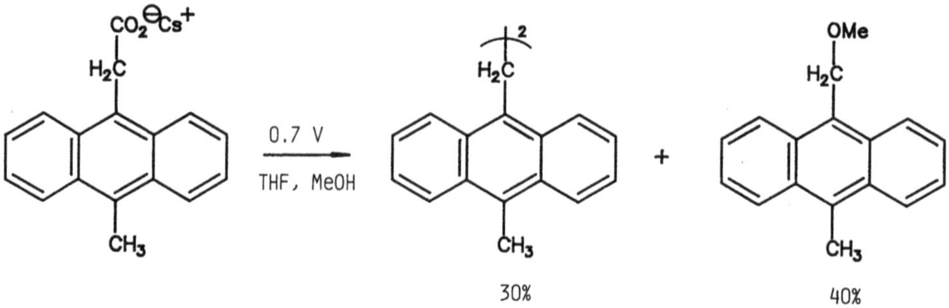

$$\text{(40)}$$

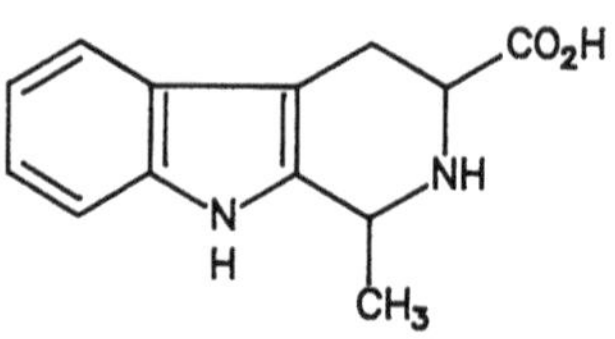

54

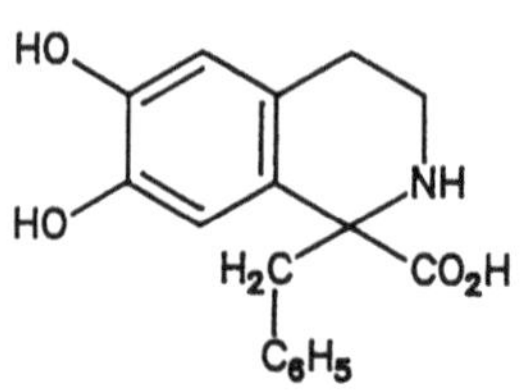

55

The decarboxylation of the caesium salt of 9-methylanthracene-10-acetic acid occurs at an even lower potential (0.7 V) and affords the dimer as well as the methyl ether (Eq. 40) [342]. The low oxidation potentials for the decarboxylation of **54** (0.13 to 0.77 V) [306a] and **55** (−0.17 V) [306b] indicate too, that the initial electron transfer occurs from the amino or aryl group rather than from the carboxylate.

14 The Photo-Kolbe Reaction

$$RCO_2^\ominus \longrightarrow RCO_2^\circ + e^\ominus \longrightarrow R^\circ + CO_2 \xrightarrow{+e^\ominus} R^\ominus \xrightarrow{H^\oplus} RH \qquad (41)$$

$$\downarrow$$

$$1/2\ R\text{---}R$$

The photo-Kolbe reaction is the decarboxylation of carboxylic acids at low voltage under irradiation at semiconductor anodes (TiO_2), that are partially doped with metals, e.g. platinum [343, 344]. On semiconductor powders the dominant product is a hydrocarbon by substitution of the carboxylate group for hydrogen (Eq. 41), whereas on an n-TiO_2 single crystal in the oxidation of acetic acid the formation of ethane besides methane could be observed [345, 346]. Dependent on the kind of semiconductor, the adsorbed metal, and the pH of the solution the extent of alkyl coupling versus reduction to the hydrocarbon can be controlled to some extent [346]. The intermediacy of alkyl radicals has been demonstrated by ESR-spectroscopy [347], that of the alkyl anion by deuterium incorporation [344]. With vicinal diacids the mono- or bisdecarboxylation can be controlled by the light flux [348]. Adipic acid yielded butane [349] with levulinic acid the products of decarboxylation, methyl ethyl-ketone, and of CC-bond cleavage, propionic acid, acetic acid, acetone, acetaldehyde were found. These can subsequently undergo decarboxylation to ethane and methane. The product distribution is a complex function of the phase of n-TiO_2 and the level of metalization of the semiconductor powder [350]. Preparative applications of the method have not been reported yet. The reaction seems attractive for several reasons. It needs only a low discharge potential, sun light is sufficient for irradiation and the carboxylgroup can be substituted for hydrogen. The latter cannot be achieved by Kolbe electrolysis and needs more steps, if the conversion is done by chemical methods. However, up to now the yields in photo-Kolbe electrolysis are very low.

15 Anodic Oxidation of Carboxylic Acids Without Decarboxylation

In carboxylic acids with an aromatic group or a double bond the π-systems can be oxidized to radical cations that react with the carboxyl group to lactones (Eqs. 7, 42) [142, 351].

In solvents that strongly resist anodic oxidation as MeCN, CH_2Cl_2/CF_3CO_2H, or FSO_3H CH-bonds in the alkyl chain can be oxidized. In acetonitrile a preferential acetamidation in the (ω-2)- and (ω-1)-position occurs (Eq. 43) [352].

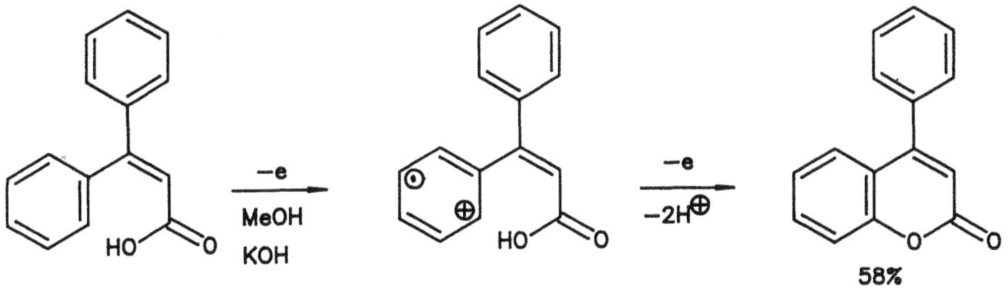

$CH_3(CH_2)_4CO_2Me$ $\xrightarrow{-e, MeCN}_{LiClO_4}$ $CH_3\overset{NHCOCH_3}{\underset{|}{CH}}$—$(CH_2)_3CO_2Me$ + CH_3CH_2—$\overset{}{\underset{NHCOCH_3}{CH}}$—$(CH_2)_2CO_2Me$

42%   11%

(43)

$CH_3(CH_2)_{12}CO_2H$ $\xrightarrow{-e, TBABF_4}_{CH_2Cl_2/CF_3CO_2H}$ $CH_3(CH_2)_x\overset{O_2CCF_3}{\underset{|}{CH}}(CH_2)_yCO_2H$ (44)

40%

y	6	7	8	9	10	11
rel. yield (%)	7	19	28	20	16	7

In methylenechloride/trifluoroacetic acid or fluorsulfonic acid trifluoracetoxylation (Eq. 44) [353] or fluorsulfonation [354] of the alkyl chain remote from the protonated carboxyl group occurs.

16 Conclusions

Carboxylic acids can be converted by anodic oxidation into radicals and/or carbocations. The procedure is simple, an undivided beaker-type cell to perform the reaction, current control, and usually methanol as solvent is sufficient. A scale up is fairly easy and the yields are generally good. The pathway towards either radicals or carbocations can be efficiently controlled by the reaction conditions (electrode material, solvent, additives) and the structure of the carboxylic acids. A broad variety of starting compounds is easily and inexpensively available from natural and petrochemical sources, or by highly developed procedures for the synthesis of carboxylic acids.

By the radical pathway $1,n$-diesters, -diketones, -dienes or -dihalides, chiral intermediates for synthesis, pheromones and unusual hydrocarbons or fatty acids are accessible in one to few steps. The addition of the intermediate radicals to double bonds affords additive dimers, whereby four units can be coupled in one step. By way of intramolecular addition unsaturated carboxylic acids can be converted into five membered hetero- or carbocyclic compounds. These radical reactions are attractive for synthesis because they can tolerate polar functional groups without protection.

Possibly the use of fatty acids as renewable resources and alternative to petrochemical feed stocks can profit from the application of Kolbe electrolysis.

The cationic pathway allows the conversion of carboxylic acids into ethers, acetals or amides. From α-aminoacids versatile chiral building blocks are accessible. The eliminative decarboxylation of vicinal diacids or β-silyl carboxylic acids, combined with cycloaddition reactions, allows the efficient construction of cyclobutenes or cyclohexadienes. The induction of cationic rearrangements or fragmentations is a potent way to specifically substituted cyclopentanoids and ring extensions by one- or four carbons. In view of these favorable qualities of Kolbe electrolysis, numerous useful applications of this old reaction can be expected in the future.

17 Acknowledgement

My own contributions to this field rely heavily on the creative cooperation with enthusiastic coworkers who are cited in the references. I am indebted to generous financial support of our work by the Arbeitsgemeinschaft Industrieller Forschungsvereinigungen e. V., the Deutsche Forschungsgemeinschaft and the Fonds der Chemischen Industrie. Finally but not least, I am thankful to Mrs. Kölle and Mrs. Quiller, who typed the manuscript with care and patience.

18 References

1. Faraday M (1834) Pogg. Ann. 33: 438
2. Kolbe H (1849) Ann. 69: 257; (1860) 113; 125; (1874) J. Prakt. Chem. [2] 10: 89; (1875) 11: 24
3. Wurtz A (1855) Ann. Chim. Phys. 44: 291
4. Brown AC, Walker J (1891) Ann. Chem. 261: 107
5. Hofer H, Moest M (1902) Ann. Chem. 323: 284
6. Fichter F (1942) Organische Elektrochemie, Steinkopff, Dresden
7. Weedon BCL (1960) Adv. Org. Chem 1: 1
8. a) Eberson L (1967) Electrochim. Acta 12: 1473
   b) (1968) Acta Chem. Scand. 22: 2462
   c) Eberson L (1963) Acta Chem. Scand. 17: 2004
9. Utley JHP (1974) In: Weinberg NL (ed) Technique of electroorganic synthesis, Wiley, vol. 5, Part 1, p 793
10. Vijh AK, Conway BE (1967) Chem. Rev. 67: 623
11. a) Beck F, Nohe H, Haufe J, Pat DE 2.023.080, BASF; (1972) CA 76: 45764n
    b) Eisele W, Nohe H, Suter H, Pat DE 2.039.991, BASF; (1972) CA 76: 120912q
    c) Beck F, Himmele W, Haufe J, Brunold A, Pat DE 1.643.693, BASF; (1977) CA 86: 54997
    d) Beck F (1973) Electrochim. Acta 18: 359
    e) Wenisch F, Nohe H, Suter H, Pat DE 2.248.562, BASF; (1974) CA 81: 32626x
12. a) Vassiliev YB, Kovsman EP, Freidlin GN (1982) Electrochim. Acta 27: 953
    b) Vassiliev YB, Kovsman EP, Freidlin GN (1982) Electrochim. Acta 27: 937
13. a) Isoya T, Kakuta R, DE 2404359; Asahi Chem.; (1974) CA 81: 135475b
    b) Isoya T, Kakuta R, Kawamura C, DE 2404560; Asahi Chem.; (1975) CA 82: 49312k
    c) Yamataka K, Matsuoka Y, Isoya T, DE 2830144; Asahi Chem.; (1979) CA 90: 112142n
    d) Yamataka K, Isoya T, Kawamura C, DE 3019537; Asahi Chem.; (1981) CA 94: 73666x
14. Allen MJ (1958) Organic electrode processes, Reinhold, New York
15. Brockmann CJ (1926) Electroorganic chemistry, Wiley, New York, pp 23–78
16. Hickling A (1949) Q. Rev., Chem. Soc. 3: 95
17. Weinberg NL, Weinberg HR (1968) Chem. Rev. 68: 449
18. Eberson L (1969) In: Patai S (ed) Chemistry of carboxylic acids and esters, Interscience, 1969, p 53
19. Eberson L, Utley JHP (1983) In: Baizer MM (ed) Organic electrochemistry, 2nd edn, Dekker, 1983, New York, p 435
20. Ross SD, Finkelstein M, Rudd EF (1975) Anodic oxidation, Academic Press, Orlando London, pp 134, 156
21. Eberson L, Nyberg K (1978) In: Bard AJ, Lund H (eds) Encyclopedia of electrochemistry of the elements, Dekker, 1978, New York, vol 12, p 261
22. Svadkovskaya GE, Voitkevich SA (1961) Russ. Chem. Rev. 29: 161
23. Schäfer HJ (1979) Chem. Phys. Lipids 24: 321
24. Baizer MM (1984) Tetrahedron 40: 935
25. Torii S (1985) Electroorganic syntheses, methods and applications, Pt. 1: Oxidation, VCH Publishers, Deerfield Beach, p 51
26. Schäfer HJ (1987) Recent advances in electroorganic synthesis, In: Torii S (ed) Proc. 1st Internat. Symposium on Electroorganic Synthesis, Kodansha, Tokyo, p 3
27. Schäfer HJ (1989) Dechema Monographie, VCH, Weinheim, vol 112, p 399
28. Dickinson T, Wynne-Jones WFK (1962) Trans. Faraday Soc. 58: 382, 388, 400
29. Ref. [18] p 58
30. Ref. [9] p 887
31. Coleman JP, Lines R, Utley JHP, Weedon BCL (1974) J. Chem. Soc., Perkin Trans. II: 1064
32. Haufe J, Beck F (1970) Chem.-Ing.-Techn. 42: 170
33. DOS 1 802 865 (1968) BASF; (1971) CA 74: 37822
34. Beck F, Nohe H, Haufe J, BASF, Germ. Offen. Nr. 2014,985; (1972) CA 76: 3402
35. a) Beck F (1973) Electrochim. Acta 18: 359
    b) Wenisch F, Nohe H, Hannebaum H, Horn RK, Stroezel M, Degner D (1979) AIChE Sym. Ser. 75: 14; (1979) CA 91: 65193
    c) DOS 2014985 (1970) BASF; (1972) CA 76: 3402 n

36. Dinh-Nguyen N (1958) Acta Chem. Scand. 12: 585
37. Sanderson JE, Levy PF, Cheng LK, Barnard GW (1983) J. Electrochem. Soc. 130: 1844
38. a) Woolford RG (1962) Can. J. Chem. 40: 1846
    b) Woolford RG, Arbic W, Rosser A (1964) ibid 42: 1788
39. Fairweather DA, Walker OJ (1926) J. Chem. Soc. 3111
40. Glasstone S, Hickling A (1939) Chem. Rev. 25: 407
41. a) Avrutskaya IA, Fioshin MY (1966) Elektrokhimiya 2: 920; (1966) CA 65: 14838e
    b) Fioshin MY, Avrutskaya IA (1967) Elektrokhimiya 3: 1288; (1968) CA 68: 26297f
42. Fioshin MY, Vasilev YB, Gaginkina EG (1960) Doklady Akad. Nauk S.S.S.R. 135: 909; (1962) CA 56: 2273d
43. a) Swann S Jr. (1956) In: Weissberger A (ed) Technique of organic chemistry, 2nd edn, vol. II, Catalytic, Photochemical and Electrolytic Reactions, Wiley-Interscience, New York, p 385
    b) Renaud RN, Sullivan DE (1973) Can. J. Chem. 51: 772
44. Adamov AA, Kovsman EP, Freidlin GN, Tarkhanov GA (1975) Elektrokhimiya 11: 1773; (1976) CA 84: 58558
45. Garwood RF, Naser-ud-Din, Scott CJ, Weedon BCL (1973) J. Chem. Soc., Perkin Trans. I: 2714
46. Rodewald LB, Lewis MC (1971) Tetrahedron 27: 5273
47. Kanevskii LS, Vasilev YB (1976) Elektrokhimiya 12: 1487; (1977) CA 86: 80805
48. Brennan MPJ, Brettle R (1973) J. Chem. Soc., Perkin Trans. I: 257
49. a) Linstead RP, Shephard BR, Weedon BCL (1951) J. Chem. Soc. 2854
    b) ibid. (1952) J. Chem. Soc. 3624
50. Rand L, Mohar AF (1965) J. Org. Chem. 30: 3156, 3885
51. a) Rudd EJ, Finkelstein M, Ross S (1972) J. Org. Chem. 37: 1763
    b) Bélanger G, Lamarre C, Vijh AK (1975) J. Electrochem. Soc. 122: 46
52. Muck DL, Wilson ER (1970) J. Electrochem. Soc. 117: 1358
53. Hawkes GE, Utley JHP, Yates GB (1976) J. Chem. Soc., Perkin Trans. II: 1709
54. Laurent A, Laurent E, Thomalla M (1972) C.R. Acad. Sci. Paris, Ser. C, 274: 1537
55. Gurjar VG (1978) J. Appl. Electrochem. 8: 207; (1978) CA 89: 82038t
56. Beck F (1974) Elektroorganische Chemie, VCH, Weinheim, Table 28, p 229
57. Tyurin JM, Kovsman EP, Karajewa EA (1965) Zh. Prikl. Khim. 38: 1818; (1966) CA 64: 3045
58. Sato N, Sekine T, Sugino K (1968) J. Electrochem. Soc. 115: 242
59. Tourillon G, Dubois JE, Lacaze PC (1977) J. Chim. Phys. Phys.-Chim. Biol. 74: 685; (1977) CA 87: 174777v
60. Yoshizawa S, Takehara Z, Ogumi Z, Matsubara M (1975) Denki Kagaku 43: 526; (1976) CA 84: 81627
61. Ogumi Z, Yamashita H, Nishio K, Takehara Z, Yoshizawa S (1983) Electrochim. Acta 28: 1687
62. Kunugi A, Urata H, Nagaura S (1968) Denki Kagaku 36: 237; (1968) CA 69: 82878
63. a) Iwakura C, Goto F, Tamura H (1980) Denki Kagaku 48: 21; (1980) CA 93: 15655m
    b) Mirkind LA, Al'bertinskii GL (1980) Elektrokhimiya 16: 911; (1980) CA 93: 83496
    c) Mirkind LA, Al'bertinskii GL, Kornienko AG (1983) Elektrokhimiya 19: 122; (1983) CA 98: 115648
64. Walker OJ, Weiss J (1935) Trans. Faraday Soc. 31: 1011 and earlier papers
65. a) Glasstone S, Hickling A (1939) Chem. Rev. 25: 407
    b) Hickling A (1949) Quart. Rev. (London) 3: 95
66. Schall C (1896) Z. Elektrochem. 3: 83
67. Ref. [9] p 796
68. Nadebaum PR, Fahidy TZ (1972) Electrochim. Acta 17: 1659
69. Ref. [20] p 139
70. Ref. [19] p 442
71. Ref. [9] p 886
72. Ref. [56] p 227
73. a) Fleischmann M, Mansfield JR, Wynne-Jones WFK (1965) J. Electroanal. Chem. 10: 511, 522
    b) Fleischmann M, Goodridge F (1968) Discuss. Faraday Soc. 45: 254
74. a) Wilson CL, Lippincott WT (1956) J. Am. Chem. Soc. 78: 4290
    b) Hickling A, Wilkins R (1986) Discuss. Faraday Soc. 45: 261; (1968) CA 69: 112843h

75. Koch DFA, Woods R (1968) Electrochim. Acta 13: 2101
76. Cervino RM, Triaca WE, Arvia AJ (1984) J. Electroanal. Chem. 172: 255
77. Conway BE, Vijh AK (1967) J. Phys. Chem. 71: 3637, 3655
78. a) Conway BE, Dzieciuch M (1963) Can. J. Chem. 41: 21, 38, 55
    b) Fairweather DA, Walker OJ (1931) Trans. Faraday Soc. 27: 722
79. a) Dubinin AG, Mirkind LA, Kazarınov VE, Fioshin MY (1979) Elektrokhimiya 15: 1337; (1979) CA 91: 165527
    b) Yakovleva AA, Kaidalova SN, Skuratnik YB, Veselovskii VI (1972) Elektrokhimiya 8: 1799; (1973) CA 78: 78958
80. Shaw JW, Nonaka T, Chou TC (1985) J. Chin. Inst. Chem. Eng. 16: 245; (1986) CA 104: 12183r
81. a) Pande GS, Shukla SN (1961) Elektrochim. Acta 4: 215
    b) Fioshin MY, Vasilev YB (1963) Izvest. Akad. Nauk SSR 437; (1963) CA 59: 8359f
    c) Dickinson T, Wynne-Jones WFK (1962) Trans. Faraday Soc. 58: 382; (1962) CA 57: 10925e
    d) Vijh AK, Conway BE (1967) Z. Anal. Chem. 230: 81
82. Sugino K, Sekine T, Sato N (1963) Electrochem. Technol. 1: 112; (1963) CA 58: 10972h
83. Girina GP, Fioshin MY, Kazarinov VE (1965) Elektrokhimiya 1: 478; (1963) CA 63: 9453h
84. a) Conway BE, Dzieciuch M (1962) Proc. Chem. Soc. 121; (1963) CA 61: 308b
    b) Conway BE, Dzieciuch M (1963) Can. J. Chem. 41: 38, 55
85. Conway BE, Liu TC (1988) J. Electroanal. Chem. 242: 317
86. Bewick A, Brown DJ (1976) Electrochim. Acta 21: 979
87. Bruno F, Dubois JE (1972) Electrochim. Acta 17: 1161
88. Russell CD (1976) J. Electroanal. Chem. 71: 81
89. Gilbert BC, Holmes RGG, Marshall PDR, Norman ROC (1977) J. Chem. Res. (S) 172
90. Sasaki K, Uneyama K, Nagaura S (1966) Electrochim. Acta 11: 891
91. Conway BE, Vijh AK (1967) Electrochim. Acta 12: 102
92. Pistorius R, Schäfer HJ: unpublished results
93. Eberson L, Ryde-Petterson G (1973) Acta Chem. Scand. 27: 1159
94. Eberson L, Nyberg K, Servin R (1976) Acta Chem. Scand. B30: 906
95. Hawkes GE, Utley JHP, Yates GB (1976) J. Chem. Soc., Perkin Trans. II: 1709
96. Utley JHP, Yates GB (1978) J. Chem. Soc., Perkin Trans. II, 395
97. Barry JE, Finkelstein M, Mayeda EA, Ross SD (1976) J. Am. Chem. Soc. 98: 8098
98. Ref. [10] p 629
99. Eberson L (1959) Acta Chem. Scand. 13: 40
100. Eberson L, Sandberg B (1966) Acta Chem. Scand. 20: 739
101. Eberson L (1960) Acta Chem. Scand. 14: 641
102. Eberson L, Nilsson S (1968) Acta Chem. Scand. 22: 2453
103. Eberson L (1963) Acta Chem. Scand. 17: 1196
104. a) Lelandais D, Chkir M (1974) Tetrahedron Lett. 36: 3113
     b) Chkir M, Lelandais D, Bacquet C (1981) Can. J. Chem. 59: 945
105. Motoki O (1955) J. Chem. Soc. Jap. 76: 930
106. Tsuji J, Kaito M, Yamada T, Mandai T (1978) Bull. Chem. Soc. Jap. 51: 1915
107. Woolford RG, Soong J, Lin WS (1967) Can. J. Chem. 45: 1837
108. Conway BE, Dzieciuch M (1963) Can. J. Chem. 41: 21
109. Waefler JP, Tissot P (1978) Electrochim. Acta 23: 899
110. Pattison FLM, Stothers JB, Woolford RG (1956) J. Am. Chem. Soc. 78: 2255
111. Fioshin MY, Tomilov AP, Avrutskaya IA, Kazakova LI, Eskin NT, Gromova GA (1963) Zh. Vses. Khim. 8: 600; (1964) CA 60: 3992h
112. Saotome K, Komoto H, Yamazaki T (1966) Bull. Chem. Soc. Jap. 39: 480
113. Berenblit VV, Panitkova ES, Rondarev DS, Sass VP, Sokolov SV (1974) Zh. Prikl. Khim. 47: 2427; (1975) CA 82: 78455
114. Levin AI, Chechina ON, Sokolov SV (1965) Zh. Obshch. Khim. 35: 1778; (1966) CA 64: 1943a
115. Coe PL, Sellers SF, Tatlow JC, Fielding HC, Wittaker G (1983) J. Chem. Soc., Perkin Trans. I: 1957
116. Kobayashi Y, Chiba T, Yokota K, Takata Y (1979) Hokkaido Daigaku Kogakubu Kenkyu Hokoku 45; (1980) CA 92: 163547
117. Mori K (1961) Nippon Kagaku Zasshi 82: 1375; (1962) CA 57: 14929e
118. Rand L, Mohar AF (1965) J. Org. Chem. 30: 3885

119. Bunyan PJ, Hey DM (1962) J. Chem. Soc. 1360
120. Stork G, Meisels A, Davies JE (1963) J. Am. Chem. Soc. 85: 3419
121. Corey EJ, Sauers RR (1959) J. Am. Chem. Soc. 81: 1739
122. Rabjohn N, Flash GW (1981) J. Org. Chem. 46: 4082
123. Kobayashi S, Jap. Patent 7686401; (1977) CA 86: 120766b
124. Zhang L, Pan X, Zhao M, Tang H, Xu B (1988) Huaxue Tongbao 41; (1988) CA 108: 212438x
125. Kimura K, Horie S, Minato I, Odaira Y (1973) Chem. Lett. 1209; see also: Binns TD, Brettle R, Cox GB (1968) J. Chem. Soc. [London] C 584
126. For a good summary see ref. [56] p 233
127. Kuei W, Yingyong (1986) Kexue Xuebao 4: 351; (1987) CA 106: 92522
128. Degner D (1988) Top. Curr. Chem. 148: 1
129. Zaidenberg AZ, Skundin AM, Vasilev YB, Kalinin YK, Dyukkiev EF, Kazarinov VE, Grinberg VA, Pat. SU 1.091.504, AS USSR; (1985) CA 103: 123023n
130. Quast H, Christ J (1984) Liebigs Ann. Chem. 1180
131. Merserau JM, Pat. DE 2.022.341, Uniroyal; (1971) CA 74: 54404f; see also: Gozlan A, Zilkha A (1984) Eur. Polym. J. 20: 759, 1199; (1984) CA 101: 152443p; (1985) 102: 114066f. For oligomers see: Adamov AA, Kovsman EP, Freidlin GN, Tarkhanov GA (1975) Elektrokhimiya 11: 1773; (1976) CA 84: 58558
132. Toy SM (1967) J. Electrochem. Soc. 114: 1042
133. Cauquis G, Haemmerle B (1970) Bull. Soc. Chim. Fr. 183
134. Wohl A, Schweitzer H (1906) Chem. Ber. 39: 890
135. Chechina ON, Levin AI (1971) Zh. Prikl. Khim. 44: 359; (1971) CA 74: 111821r
136. Knolle J, Schäfer HJ (1978) Electrochim. Acta 23: 5
137. Yadav AK, Jain A, Misra RA (1982) Electrochim. Acta 27: 535
138. a) Garwood RF, Naser-ud-Din, Scott CJ, Weedon BCL (1973) J. Chem. Soc., Perkin Trans. I: 2714
    b) Knolle J, Schäfer HJ: unpublished results
139. Baldwin JE (1976) J. Chem. Soc., Chem. Commun. 734
140. Huhtasaari M, Schäfer HJ, Luftmann H (1983) Acta Chem. Scand. B37: 537
141. Mandell L, Daley RF, Day Jr. RA (1976) J. Org. Chem. 41: 4087
142. Adams C, Jacobsen N, Utley JHP (1978) J. Chem. Soc., Perkin Trans. II: 1071
143. Bounds DG, Linstead RP, Weedon BCL (1953) J. Chem. Soc. 2393
144. Weiper A, Schäfer HJ: unpublished results
145. Matsuda Y, Kimura K, Iwakura C, Tamura H (1973) Bull. Chem. Soc. Jap. 46: 430
146. a) Greaves WS, Linstead RP, Shephard BR, Thomas SLS, Weedon BCL (1950) J. Chem. Soc. 3326
147. Petersen RC, Finkelstein M, Ross SD (1967) J. Org. Chem. 32: 564
148. E.g. perfluornonanoic acid and methyl azelate: Weiper A, Schäfer HJ: unpublished results
149. Renaud RN, Sullivan DE (1972) Can. J. Chem. 50: 3084
150. a) Feldhues M, Schäfer HJ (1985) Tetrahedron 41: 4195, 4213; (1986) ibid. 42: 1285
    b) Lomölder R, Schäfer HJ (1987) Angew. Chem. 99: 1282; (1987) Angew. Chem. Int. Ed. Engl. 26: 1253
151. Baker BW, Kierstead RW, Linstead RP, Weedon BCL (1954) J. Chem. Soc. 1804
152. Krasavtsev II (1980) Ukr. Khim. Zh. 46: 776; (1980) CA 93: 139974j
153. Naser-ud-Din (1976) Pak. J. Sci. Ind. Res. 19: 132; (1978) CA 88: 120537
154. Rawlings FF (1964) J. Electrochem. Techn. 2: 217; (1964) CA 61: 9172
155. Brettle R, Latham DW (1968) J. Chem. Soc. (C) 906
156. Motoki K, Odoka S (1956) J. Chem. Soc. Jap., Pure Chem. Sect. 77: 163
157. Fuchs W, Barnetzky E (1955) Fette u. Seifen 57: 675
158. Ställberg-Stenhagen S (1951) Ark. Kemi 2: 95
159. Mislow K, Steinberg IV (1955) J. Am. Chem. Soc. 77: 3807
160. Romanuk M, Streinz L, Sorm F (1972) Collect. Czech. Chem. Commun. 37: 1755
161. Odham G, Petterson B, Stenhagen E (1974) Acta Chem. Scand. (B) 28: 36
162. Linstead RP, Lunt JC, Weedon BCL (1951) J. Chem. Soc. 1130
163. a) Tember GA, Getmanskaya ZI, Nichikova PR (1974) Zh. Prikl. Khim. 47: 477; (1974) CA 80: 120168j

b) Taikov BF, Novakovskii EM, Zhelhovskaya VP, Shadrova VN, Shcherbik PK (1981) Khim. Tverd. Topl. 61; (1982) CA 96: 51767v
c) Jap. Pat. (1985) CA 102: 228323t
164. Eberson L (1962) J. Org. Chem. 27: 3706
165. Okida Y, Jap. Pat., Okamura Oil Mill; (1987) CA 106: 127947
166. a) Takahashi M, Osawa K, Ueda J, Okada K (1976) Yakugaku Zasshi 96: 1000; (1976) CA 84: 135228k
b) Schäfer HJ, Jensen U: unpublished results
c) Naser-ud-Din (1976) Pak. J. Sci. Ind. Res. 19: 132; (1978) CA 88: 120571
167. Seidel W, Schäfer HJ (1980) Chem. Ber. 113: 3898
168. Ref. [9] Table 6.11
169. Berenblit VV, Zapevalov AY, Panitkova ES, Plashkin VS, Rondarev DS, Sass VP, Sokolov SV (1979) Zh. Org. Khim. 15: 1417; (1979) CA 91: 192779
170. Renaud RN, Champagne PJ (1975) Can. J. Chem. 53: 529
171. Suhura Y, Mijazaki S (1969) Bull. Chem. Soc. Jap. 42: 3022
172. Andreev VM, Polyakova SG, Khrustova ZS, Bazhulina V, Smirnova VV (1980) USSR Pat.; (1981) CA 95: 6513c
173. Schäfer HJ, Hermeling D: unpublished results
174. Jap. Pat. 57.200.576, Asahi Glass Co; (1983) CA 98: 206529
175. Jap. Pat. 58.161.784 Asahi Chem. Industry Co; (1984) CA 100: 128859
176. Stoll M (1951) Helv. Chim. Acta 34: 1817
177. Motoki S, Yamada Y (1960) Nippon Kagaku Zasshi 81: 665; (1962) CA 56: 396i
178. Andreev VM, Polyakova SG, Bazhulina VI, Khrustova ZS, Smirnova VV, Gorbunkova VP, Shchedrina MM (1981) Zhur. Organ. Khim. 17: 86; (1981) CA 94: 208308c
179. Levy PF, Sanderson JE, Cheng LK (1984) J. Electrochem. Soc. 131: 773
180. Linstead RP, Shephard BR, Weedon BCL, Lunt LC (1953) J. Chem. Soc. 1538
181. a) Linstead RP, Weedon BCL, Wladislaw B (1955) J. Chem. Soc. 1097
b) Bounds DG, Linstead RP, Weedon BCL (1954) J. Chem. Soc. 4219
182. Kihira K, Batta A, Mosbach EH, Salen G (1979) J. Lipid Res. 20: 421; (1979) CA 91: 175606
183. a) Baker BW, Linstead RP, Weedon BCL (1955) J. Chem. Soc. 2218
b) Takahashi M Jap. Pat. 76.113.835, Sanei Chem. Ind.; (1977) CA 86: 155359
184. Linstead RP, Lunt JW, Weedon BCL, Shephard BR (1952) J. Chem. Soc. 3621
185. Takahashi M Jap. Pat. 76.113.836, Sanei Chem. Ind.; (1977) CA 86: 139611
186. Bounds DG, Linstead RP, Weedon BCL (1954) J. Chem. Soc. 448
187. Pattison FLM, Woolford RG (1957) J. Am. Chem. Soc. 79: 2306, 2308
188. Gribble GW, Sanstead JK, Sullivan JW (1973) J. Chem. Soc., Chem. Commun. 735
189. Seidel W, Schäfer HJ: unpublished results
190. Yadav AK, Tissot P (1984) Helv. Chim. Acta 67: 1698
191. Klünenberg H, Schäfer HJ (1978) Angew. Chem. 90: 48; (1978) Angew. Chem. Int. Ed. Engl. 17: 47
192. Jensen U, Schäfer HJ (1981) Chem. Ber. 114: 292
193. Bestmann HJ, Roth K, Michaelis K, Vostrowsky O, Schäfer HJ, Michaelis R (1987) Liebigs Ann. Chem. 417
194. Seidel W, Schäfer HJ (1980) Chem. Ber. 113: 451; see also: Jensen-Korte U, Schäfer HJ (1982) Liebigs Ann. Chem. 1532
195. a) Linstead RP, Lunt JC, Weedon BCL (1950) J. Chem. Soc. 3331
b) Linstead RP, Lunt JC, Weedon BCL (1951) J. Chem. Soc. 1130
196. Takahashi J, Mori K, Matsui M (1979) Agric. Biol. Chem. 43: 1605; (1979) CA 91: 174752
197. Seidel W, Knolle J, Schäfer HJ (1977) Chem. Ber. 110: 3544
198. Schäfer HJ, Knolle J: unpublished results
199. Schäfer HJ, Wittenbrink C: unpublished results
200. Knolle J, Schäfer HJ (1975) Angew. Chem. 87: 777; (1975) Angew. Chem. Int. Ed. Engl. 14: 758
201. a) Chechina ON, Bildinov KN, Levin AI, Sokolov SV, USSR Pat. 370.199; (1973) CA 79: 18135
b) Chechina ON, Levin AI (1974) Elektrokhimiya 10: 1170; (1974) CA 81: 113916

202. Nutt RF, Strachan RG, Veber DF, Holly FW (1980) J. Org. Chem. 45: 3078
203. Seebach D, Renaud P (1985) Helv. Chim. Acta 68: 2342
204. Ceder O, Nilsson HG (1977) Acta Chem. Scand. B31: 189
205. Smets G, van Borght X, van Haeren G (1964) J. Polym. Sci. Pt. A 2: 5187; (1965) CA 62: 12751a
206. Schäfer H, Pistorius R (1972) Angew. Chem. 84: 893; (1972) Angew. Chem. Int. Ed. Engl. 11: 841
207. Lindsey RV, Peterson ML (1959) J. Am. Chem. Soc. 81: 2073
208. Fioshin MY, Kamneva LA, Mirkind LA, Salmin LA, Kornienko AG (1962) Neftekhimiya 2: 557; (1963) CA 58: 11205b
209. Fioshin MY, Kamnewa AI, Mirkind LA, Kornienko AG, Salmin LA (1966) Khim. Prom. 42: 804; (1967) CA 66: 43109x
210. Fioshin MY, Mirkind LA, Salmin LA, Kornienko AG (1965) Zh. Chim. Obsch. 10: 238; (1965) CA 63: 15858c
211. Nechiporenko VP, Bogoslovskii KG, Novikov NA, Guigina NI, Mirkind LA (1984) Zh. Org. Khim. 20: 1799; (1985) CA 102: 61763
212. Goldschmidt S, Stöckl E (1952) Chem. Ber. 85: 630
213. Vassiliev YB, Bagotzky VS, Kovsman EP, Grinberg VA, Kanevsky LS, Polishchyuk VR (1982) Electrochim. Acta 27: 919
214. Vassiliev YB, Lotvin BM, Grinberg VA (1981) Elektrokhimiya 17: 1252; (1981) CA 95: 211885
215. Smith WB, Hyon Yuh Y (1968) Tetrahedron 24: 1163
216. Chkir M, Lelandais D (1971) J. Chem. Soc., Chem. Commun. 1369
217. Karapetyan KG, Bezzubov AA, Kanevskii LS, Skundin AM, Vasilev YB (1976) Elektrokhimiya 12: 1623; (1977) CA 86: 62705t
218. Renaud RN, Champagne PJ (1979) Can. J. Chem. 57: 990
219. Bogoslovskii KG, Mirkind LA, Kondrikov NB (1986) Zh. Vses. Khim. 31: 470; (1986) CA 105: 215678z
220. Champagne PJ, Renaud RN (1980) Can. J. Chem. 58: 1101
221. Uneyama K, Ueda K (1988) Chem. Lett. 853
222. Stork L, Schäfer HJ: unpublished results
223. Brookes CJ, Coe PL, Owen DM, Pedler AE, Tatlow JC (1974) J. Chem. Soc., Chem. Commun. 323; see also: Brookes CJ, Coe PL, Pedler AE, Tatlow JC (1978) J. Chem. Soc., Perkin Trans. I: 202
224. Muller N (1983) J. Org. Chem. 48: 1370
225. Renaud RN, Champagne PJ, Savard M (1979) Can. J. Chem. 57: 2617
226. Renaud RN, Stephens CJ, Bérubé D (1982) Can. J. Chem. 60: 1687
227. Bruno F, Dubois JE (1973) Bull. Soc. Chim. Fr. 2270
228. Smith WB, Gilde HG (1959) J. Am. Chem. Soc. 81: 5325; (1961) J. Am. Chem. Soc. 83: 1355
229. Vassiliev YB, Bagotsky VS, Kovsman EP, Grinberg VA, Kanevsky LS, Polishchyuk VR (1982) Electrochim. Acta 27: 929
230. Buist PH, Kendall J, Barradas RG (1984) J. Electroanal. Chem. 161: 393
231. Smith WB, Gilde HG (1960) J. Am. Chem. Soc. 82: 659
232. Smith WB, Manning DT (1962) J. Polym. Sci. 59: 945; (1963) CA 55: 27003a
233. a) Ogumi Z, Tari I, Takehara Z, Yoshizawa S (1976) Bull. Chem. Soc. Jap. 49: 841, 2883
    b) Yoshizawa S, Takehara Z, Ogumi Z, Najai C (1972) Denki Kagaku 40: 724; (1973) CA 79: 19194u
234. Utley JHP, Holman RJ (1976) Electrochim. Acta 21: 987
235. Grinberg VA, Polishchuk VP, Kanevskii LS, German LS (1980) CA 93: 157798
236. Schäfer HJ, Huhtasaari M, Becking L (1984) Angew. Chem. 96: 995; (1984) Angew. Chem. Int. Ed. Engl. 23: 980
237. Becking L, Schäfer HJ (1988) Tetrahedron Lett. 29: 2797
238. Schäfer HJ, Dralle G: unpublished results
239. Becking L, Schäfer HJ (1988) Tetrahedron Lett. 29: 2801
240. Giese B (1986) Radicals in organic synthesis, Pergamon, Oxford
241. For kinetic studies see: Tyurin YM, Afonshin GN (1969) Elektrokhimiya 5: 1198; (1970) CA 72: 27687; Vijh AK (1972) J. Elektrochem. Soc. 119: 679
242. a) Walling C (1957) Free radicals in solution, Wiley, New York, p 581
    b) Overberger CG, Kabasakalian P (1957) J. Am. Chem. Soc. 79: 3182

c) Corey EJ, Bauld NL, Lalonde RT, Casanova J, Kaiser ET (1960) J. Am. Chem. Soc. 82: 2645
243. Koehl Jr. WJ (1964) J. Am. Chem. Soc. 86: 4686
244. Sato N, Sekine T, Sugino K (1968) J. Electrochem. Soc. 115: 242
245. Kase K, Sato N, Sekine T (1983) Denki Kagaku 51: 749; (1984) CA 100: 120310f
246. a) Gassman PG, Zalar FV (1966) J. Am. Chem. Soc. 88: 2252
    b) Ross SD, Finkelstein M (1969) J. Org. Chem. 34: 2923
247. Traynham JG, Green EE, Frye RL (1970) J. Org. Chem. 35: 3611
248. Zorge JA van, Strating J, Wynberg H (1970) Recl. Trav. Chim. Pays-Bas 89: 781
249. a) Shono T, Nishiguchi I, Yamane S, Oda R (1969) Tetrahedron Lett. 1965
    b) Gassman PG, Zalar FV (1966) J. Am. Chem. Soc. 88: 2252
249. Arora PC, Woolford RG (1971) Can. J. Chem. 49: 2681
250. a) Skell PS, Starer I (1959) J. Am. Chem. Soc. 81: 4117; (1962) J. Am. Chem. Soc. 84: 3962
    b) Skell PS, Reichenbacher PH (1968) J. Am. Chem. Soc. 90: 2309
251. Skell PS, Starer I, from Deno NC (1964) Chem. Eng. News 42: 88
252. Keating JT, Skell PS (1969) J. Am. Chem. Soc. 91: 695
253. Skell PS, Reichenbacher PH (1968) J. Am. Chem. Soc. 90: 3436
254. Laurent E, Milhaud J, Marquet B, Thomalla M (1981) Nouv. J. Chim. 5: 575
255. a) Stepanov FN, Baklan VF, Guts SS (1965) Akad. Nauk SSR 97: (1966) CA 65: 627
    b) Stepanov FN, Yurchenko AG, Isaeva SS, Novikova I, Novosti Elektrokhim. Org. Soedin.; (1975) CA 82: 364950d
256. Bernlöhr W, Beckhaus HD, Lindner HJ, Rüchardt C (1984) Chem. Ber. 117: 3303
257. Banda FM, Brettle R (1977) J. Chem. Soc., Perkin Trans. I: 1773
258. Wladislaw B, Ayres AMJ (1962) J. Org. Chem. 27: 281
259. Wladislaw B, Giora A (1965) J. Chem. Soc., London 5745
260. Linstead RP, Shephard BR, Weedon BCL (1952) J. Chem. Soc. (London) 3624
261. Finkelstein M, Petersen RC (1960) J. Org. Chem. 25: 136
262. a) Rabjohn N, Cranor WL, Schofield CM (1984) J. Org. Chem. 49: 1732
    b) Wladislaw B, Zimmermann JP (1970) J. Chem. Soc. B: 290
263. Iwasaki T, Nishitani T, Horikawa H, Inoue I (1982) J. Org. Chem. 47: 3799
264. Tanaka H, Kobayashi Y, Torii S (1976) J. Org. Chem. 41: 3482
265. Wladislaw B (1962) Chem. Ind. 1868; see also Uneyama K, Torii S, Oae S (1971) Bull. Chem. Soc. Jap. 44: 815
266. Schäfer HJ, Michaelis R: unpublished results
267. Wuts PGM, Sutherland C (1982) Tetrahedron Lett. 23: 3987
268. Nokami J, Kawada M, Okawara R, Torii S, Tanaka H (1979) Tetrahedron Lett. 1045
269. a) Mitzlaff M, Schnabel H, Germ. Pat. 2.336.976 (Höchst); (1975) CA 83: 9239
    b) Linstead RP, Shephard BR, Weedon BCL (1951) J. Chem. Soc. 2854
    c) Thomas HG, Kessel S (1988) Chem. Ber. 121: 1575
    d) Finkelstein M, Ross SD (1972) Tetrahedron 28: 4497
270. a) Horikawa H, Iwasaki T, Matsumoto K, Miyoshi M (1976) Tetrahedron Lett. 191
    b) Iwasaki T, Horikawa H, Matsumoto K, Miyoshi M (1979) Bull. Chem. Soc. Jap. 52: 826
271. Horikawa H, Iwasaki T, Matsumoto K, Miyoshi M (1979) J. Org. Chem. 43: 335
272. Hess U, Gross T, Thiele R (1979) Z. Chem. 19: 195
273. a) Renaud P, Seebach D (1986) Angew. Chem. 98: 836; (1986) Angew. Chem. Int. Ed. Engl. 25: 843
    b) Steckhan E, Herborn C, Papadopoulos A, Lewall B, Ginzel KD (1988) 14. Sandbjerg Meeting, Abstr. p 59
274. a) Renaud P, Seebach D (1986) Helv. Chim. Acta 69: 1704
    b) Renaud P, Seebach D (1986) Synthesis 424
275. Yoshikawa M, Kamigauchi T, Ikeda Y, Kitagawa I (1981) Chem. Pharm. Bull. 29: 2571, 2582; (1981) Heterocycles 15: 349
276. Torii S, Tanaka H, Ogo H, Yamasita S (1971) Bull. Chem. Soc. Jap. 44: 1079
277. Iwasaki T, Horikawa H, Matsumoto K, Miyoshi M (1977) J. Org. Chem. 42: 2419
278. Thomas HG, Katzer E (1974) Tetrahedron Lett. 887
279. Mandell L, Daley RF, Day RA, Jr. (1977) J. Org. Chem. 42: 1461

280. Filardo G, Di Quarto F, Gambino S, Silvestri G (1982) J. Appl. Electrochem. 12: 127; (1982) CA 96: 59806
281. Torii S, Okamoto T, Tanida G, Hino H, Kitsuya Y (1976) J. Org. Chem. 41: 166
282. Torii S, Inokuchi T, Mizoguchi K, Yamazaki M (1979) J. Org. Chem. 44: 2303
283. a) Lelandais D, Bacquet C, Einhorn J (1981) Tetrahedron 37: 3131
    b) Bacquet C, Einhorn J, Lelandais D (1980) J. Heterocycl. Chem. 17: 831
284. a) Nokami J, Matsuura M, Sueoka T, Kusumoto Y, Kawada M (1978) Chem. Lett. 1283
    b) Torii S, Tanaka H, Kobayasi Y, Nokami J, Kawata M (1979) Bull. Chem. Soc. Jap. 52: 1553
    c) Nokami J, Kawada M, Okawara R, Torii S, Tanaka H (1979) Tetrahedron Lett. 1045
    d) Nokami J, Yamamoto T, Kawada M, Izumi M, Ochi N, Okawara R (1979) Tetrahedron Lett. 1047
285. Renaud P, Hürzeler M, Seebach D (1987) Helv. Chim. Acta 70: 292
286. Renaud P, Seebach D (1986) Helv. Chim. Acta 69: 1704
287. Iwasaki T, Horikawa H, Matsumoto K, Miyoshi M (1979) J. Org. Chem. 44: 1552
288. Nishitani T, Iwasaki T, Mushika Y, Miyoshi M (1979) J. Org. Chem. 44: 2019
289. Nishitani T, Horikawa H, Iwasaki T, Matsumoto K, Inoue I, Miyoshi M (1982) J. Org. Chem. 47: 1706
290. Mori M, Kagechika K, Tohjima K, Shibasaki M (1988) Tetrahedron Lett. 29: 1409
291. Thomas HG, Kessel S (1985) Chem. Ber. 118: 2777; (1986) Chem. Ber. 119: 2173
292. Thomas HG, Katzer K (1974) Tetrahedron Lett. 887
293. Ref. [25] p 60
294. a) Eberson L, Nyberg K (1964) Acta Chem. Scand. 18: 1567
    b) Eberson L, Olofsson B (1969) Acta Chem. Scand. 23: 2355
295. a) Kornprobst JM, Laurent A, Laurent-Dieuzeide E (1968) Bull. Soc. Chim. Fr. 3657
    b) Kornprobst JM, Laurent A, Laurent-Dieuzeide E (1970) Bull. Soc. Chim. Fr. 1490
296. a) Thomas HG (1971) Angew. Chem. 83: 579; (1971) Angew. Chem. Int. Ed. Engl. 10: 557
    b) Thomas HG (1975) Chem. Ber. 108: 967
297. Laurent E, Thomalla M (1977) Bull. Soc. Chim. Fr. 834
298. a) Laurent E, Thomalla M (1976) Tetrahedron Lett. 4727
    b) Laurent E, Thomalla M (1975) Tetrahedron Lett. 4411
    c) Laurent E, Thomalla M, Marquet B, Burger U (1980) J. Org. Chem. 45: 4193
299. Laurent E, Thomalla M (1977) Electrochim. Acta 22: 531
300. Inayama S, Kawamata T, Shimizu N (1980) Chem. Pharm. Bull. 28: 277, 282; (1980) CA 93: 25868, 168426n
301. Torii S, Okamoto T, Tanaka H (1974) J. Org. Chem. 39: 2486
302. Slobbe J (1977) Chem. Commun. 82
303. Lelandais D, Chkir M (1975) C.R. Acad. Sci. Paris, Ser. C, 281: 731
304. Diaz A (1977) J. Org. Chem. 42: 3949
305. Iwasaki T, Horikawa H, Matsumoto K, Miyoshi M (1978) Tetrahedron Lett. 4799
306. a) Bobbitt JM, Willis JP (1980) J. Org. Chem. 45: 1978
    b) Bobbitt JM, Cheng TY (1976) J. Org. Chem. 41: 443
307. Shono T, Ohmizu H, Kise N (1980) Chem. Lett. 1517
308. Hermeling D, Schäfer HJ (1988) Chem. Ber. 121: 1151
309. Westberg HH, Dauben HJ (1968) Tetrahedron Lett. 5123
310. a) Radlick P, Klem R, Spurlock S, Sims JJ, van Tamelen EE, Whitesides T (1968) Tetrahedron Lett. 5117
    b) van Tamelen EE, Carty D (1967) J. Am. Chem. Soc. 89: 3922
311. Baker AJ, Chalmers AM, Flood WW, Mac Nicol DD, Penrose AB, Raphael RA (1970) J. Chem. Soc., Chem. Commun. 166
312. Warren CB, Bloomfield JJ (1973) J. Org. Chem. 38: 4011
313. Leftin JH, Redpath D, Pines A, Gil-Av E (1973) Isr. J. Chem. 11: 75
314. Stelzer F, Brunthaler JK, Leising G, Hummel K (1986) J. Mol. Cat. 36: 135 (1986) CA 105: 173079
315. Plieninger H, Lehnert W (1967) Chem. Ber. 100: 2427
316. Vellturo AF, Griffin GW (1965) J. Am. Chem. Soc. 87: 3021
317. Kelly RC, Schletter J (1973) J. Am. Chem. Soc. 95: 7156
318. Torii S, Okamoto T, Oida T (1978) J. Org. Chem. 43: 2294

319. a) Traynham JG, Dehn JS (1967) J. Am. Chem. Soc. 89: 2139
     b) Traynham JG, Green EE, Frye RL (1970) J. Org. Chem. 35: 3611
320. a) Snow RA, Degenhardt CR, Paquette LA (1976) Tetrahedron Lett. 4447
     b) Schäfer HJ, Burgbacher G: unpublished results
321. Corey EJ, Casanova J (1963) J. Am. Chem. Soc. 85: 165
322. Maier G, Bosslet F (1972) Tetrahedron Lett. 4483
323. Shono T, Hayashi T, Omoto H, Matsumura Y (1977) Tetrahedron Lett. 2667
324. Torii S, Tanaka H, Mandai T (1975) J. Org. Chem. 40: 2221
325. Gream GE, Pincombe CF (1974) Aust. J. Chem. 27: 589
326. Gassmann PG, Fox BL (1967) J. Org. Chem. 32: 480
327. Imagawa T, Akiyama T, Kawanisi M et al. (1978) Tetrahedron Lett. 2165; (1979) Tetrahedron Lett. 1691; (1980) J. Org. Chem. 45: 2005
328. Schäfer HJ, Müller U: unpublished results
329. Winstein S, Trifan D (1952) J. Am. Chem. Soc. 74: 1147, 1154
330. a) Takeda A, Wada S, Murakami Y (1971) Bull. Chem. Soc. Jap. 44: 2729
     b) Binns TB, Brettle R, Cox GB (1969) J. Chem. Soc. C 1227, 2499
331. Waters JA, Witkop B (1971) J. Org. Chem. 36: 3232
332. Takeda A, Moriwake T, Torii S, Takaki T (1972) Bull. Chem. Soc. Jap. 45: 3718
333. a) Lelandais D, Bacquet C, Einhorn J (1978) J. Chem. Soc., Chem. Commun. 194
     b) Lelandais D, Bacquet C, Einhorn J (1981) Tetrahedron 37: 3131
334. Bonner WA, Mango FD (1964) J. Org. Chem. 29: 430
335. Schäfer HJ, Bitenic M: unpublished results
336. Rand L, Rao CS (1968) J. Org. Chem. 33: 2704
337. Wharton PS, Hiegel GA, Coombs RV (1963) J. Org. Chem. 28: 3217
338. Michaelis R, Müller U, Schäfer HJ (1987) Angew. Chem. 99: 1049; (1987) Angew. Chem. Int. Ed. Engl. 26: 1026
339. Corey EJ, Sauers RR (1959) J. Am. Chem. Soc. 81: 1739, 1743
340. Schäfer HJ, Schenk B: unpublished results
341. Thomas HG, Gabriel J, Fleischhauer J, Raabe G (1983) Chem. Ber. 116: 375
342. Coleman JP, Eberson L (1971) J. Chem. Soc., Chem. Commun. 1300
343. Kräutler B, Bard AJ (1978) J. Am. Chem. Soc. 100: 2239
344. Ibid. 5958
345. Kräutler B, Bard AJ (1979) Nouv. J. Chim. 3: 31
346. Yoneyama H, Takao Y, Tamura H, Bard AJ (1983) J. Phys. Chem. 87: 1417
347. Kräutler B, Jaeger CD, Bard AJ (1978) J. Am. Chem. Soc. 100: 4903
348. Fox MA, Chen CC, Park KH, Younathan JN (1985) Amer. Chem. Soc. Symp. Ser. 278: 69; From Fox MA (1987) Top. Curr. Chem. 142: 71
349. Izumi I, Fan FRF, Bard AJ (1981) J. Phys. Chem. 85: 218
350. Chum HL, Ratcliff M, Posey FL, Turner JA, Nozik AJ (1983) J. Phys. Chem. 87: 3089
351. a) Koehl WJ (1967) J. Org. Chem. 32: 614
     b) Thomas HG, Schwager HW (1984) Tetrahedron Lett. 25: 4471
352. Miller LL, Ramachandran V (1974) J. Org. Chem. 39: 369
353. Hembrock A, Schäfer HJ (1985) Angew. Chem. 97: 1048; (1985) Angew. Chem. Int. Ed. Engl. 24: 1055
354. Pletcher D, Smith CZ (1975) J. Chem. Soc., Perkin Trans. I: 948

Author Index Volume 151—152

Author Index Vols. 26–50 see Vol. 50
Author Index Vols. 50–100 see Vol. 100
Author Index Vols. 101–150 see Vol. 150

The volume numbers are printed in italics

Caffrey, M.: Structural, Mesomorphic and Time-Resolved Studies of Biological Liquid Crystals and Lipid Membranes Using Synchrotron X-Radiation. *151*, 75–109 (1989).
Dartyge, E., see Fontaine, A.: *151*, 179–203 (1989).
Fontaine, A., Dartyge, E., Itie, J. P., Jucha, A., Polian, A., Tolentino, H. and Tourillon, G.: Time-Resolved X-Ray Absorption Spectroscopy Using an Energy Dispersive Optics: Strengths and Limitations. *151*, 179–203 (1989).
Fuller, W., see Greenall, R.: *151*, 31–59 (1989).
Gehrke, R.: Research on Synthetic Polymers by Means of Experimental Techniques Employing Synchrotron Radiation. *151*, 111–159 (1989).
Gislason, E. A.: see Guyon, P.-M.: *151*, 161–178 (1989).
Greenall, R., Fuller, W.: High Angle Fibre Diffraction Studies on Conformational Transitions in DNA Using Synchrotron Radiation. *151*, 31–59 (1989).
Guyon, P.-M., Gislason, E. A.: Use of Synchrotron Radiation to Study State-Selected Ion-Molecule Reactions. *151*, 161–178 (1989).
Heinze, J.: Electronically Conducting Polymers. 152, 1–48 (1989).
Helliwell, J., see Moffat, J. K.: *151*, 61–74 (1989).
Holmes, K. C.: Synchrotron Radiation as a Source for X-Ray Diffraction — The Beginning. *151*, 1–7 (1989).
Itie, J. P., see Fontaine, A.: *151*, 179–203 (1989).
Jucha, A., see Fontaine, A.: *151*, 179–203 (1989).
Lange, F., see Mandelkow, E.: *151*, 9–29 (1989).
Mandelkow, E., Lange, G., Mandelkow, E.-M.: Applications of Synchrotron Radiation to the Study of Biopolymers in Solution: Time-Resolved X-Ray Scattering of Microtubule Self-Assembly and Oscillations. *151*, 9–29 (1989).
Mandelkow, E.-M., see Mandelkow, E.: *151*, 9–29 (1989).
Merz, A.: Chemically Modified Electrodes. 152, 49–90 (1989).
Moffat, J. K., Helliwell, J.: The Laue Method and its Use in Time-Resolved Crystallography. *151*, 61–74 (1989).
Polian, A., see Fontaine, A.: *151*, 179–203 (1989).
Riekel, C.: Experimental Possibilities in Small Angle Scattering at the European Synchrotron Radiation Facility. *151*, 205–229 (1989).
Schäfer, H.-J.: Recent Contributions of Kolbe Electrolysis to Organic Synthesis. 152, 91–151 (1989).
Tolentino, H., see Fontaine, A.: *151*, 179–203 (1989).
Tourillon, G., see Fontaine, A.: *151*, 179–203 (1989).

GPSR Compliance
The European Union's (EU) General Product Safety Regulation (GPSR) is a set of rules that requires consumer products to be safe and our obligations to ensure this.

If you have any concerns about our products, you can contact us on

ProductSafety@springernature.com

In case Publisher is established outside the EU, the EU authorized representative is:

Springer Nature Customer Service Center GmbH
Europaplatz 3
69115 Heidelberg, Germany

www.ingramcontent.com/pod-product-compliance
Ingram Content Group UK Ltd.
Pitfield, Milton Keynes, MK11 3LW, UK
UKHW050410240426
12048UKWH00020B/1436